Physics Research and Technology

Physics Research and Technology

Second Harmonic Generation: Pathways of Nonlinear Photonics
Abdel-Baset M. A. Ibrahim, PhD and Pankaj Kumar Choudhury, PhD
2022. ISBN: 978-1-68507-888-1 (Hardcover)
2022. ISBN: 9 979-8-88697-002-9 (eBook)

Quantum Field Theory and Applications
Natale Palerma (Editor)
2022. ISBN: 978-1-68507-930-7 (Hardcover)
2022. ISBN: 978-1-68507-957-4 (eBook)

Spatial Homeostasis, Quantum Information Channel, and the Nature of Living Things Within the Framework of the Theory of Byuon
Yuriy Baurov, PhD (Author)
2022. ISBN: 978-1-68507-590-3 (eBook)

A Closer Look at Plasmonics
Christian Bosch (Editor)
2022. ISBN: 978-1-68507-709-9 (Hardcover)
2022. ISBN: 978-1-68507-725-9 (eBook)

An Introduction to Charge Carriers
Jai Singh, PhD (Editor)
2022. ISBN: 978-1-68507-456-2 (Hardcover)
2022. ISBN: 978-1-68507-531-6 (eBook)

Handbook of Research on Heat Transfer
Suvanjan Bhattacharyya, PhD (Editor)
Varun Goel, PhD (Editor)
2022. ISBN: 978-1-68507-459-3 (Hardcover)
2021. ISBN: 978-1-68507-542-2 (eBook)

More information about this series can be found at
https://novapublishers.com/product-category/series/physics-research-and-technology/

Riccardo Zancan and Raul Tozzi

Beyond Special Relativity

Looking for the Intrinsic Properties of Space-Time

Copyright © 2022 by Nova Science Publishers, Inc.
https://doi.org/10.52305/UOXU1773

All rights reserved. No part of this book may be reproduced, stored in a retrieval system or transmitted in any form or by any means: electronic, electrostatic, magnetic, tape, mechanical photocopying, recording or otherwise without the written permission of the Publisher.

We have partnered with Copyright Clearance Center to make it easy for you to obtain permissions to reuse content from this publication. Simply navigate to this publication's page on Nova's website and locate the "Get Permission" button below the title description. This button is linked directly to the title's permission page on copyright.com. Alternatively, you can visit copyright.com and search by title, ISBN, or ISSN.

For further questions about using the service on copyright.com, please contact:
Copyright Clearance Center
Phone: +1-(978) 750-8400   Fax: +1-(978) 750-4470   E-mail: info@copyright.com.

NOTICE TO THE READER

The Publisher has taken reasonable care in the preparation of this book, but makes no expressed or implied warranty of any kind and assumes no responsibility for any errors or omissions. No liability is assumed for incidental or consequential damages in connection with or arising out of information contained in this book. The Publisher shall not be liable for any special, consequential, or exemplary damages resulting, in whole or in part, from the readers' use of, or reliance upon, this material. Any parts of this book based on government reports are so indicated and copyright is claimed for those parts to the extent applicable to compilations of such works.

Independent verification should be sought for any data, advice or recommendations contained in this book. In addition, no responsibility is assumed by the Publisher for any injury and/or damage to persons or property arising from any methods, products, instructions, ideas or otherwise contained in this publication.

This publication is designed to provide accurate and authoritative information with regard to the subject matter covered herein. It is sold with the clear understanding that the Publisher is not engaged in rendering legal or any other professional services. If legal or any other expert assistance is required, the services of a competent person should be sought. FROM A DECLARATION OF PARTICIPANTS JOINTLY ADOPTED BY A COMMITTEE OF THE AMERICAN BAR ASSOCIATION AND A COMMITTEE OF PUBLISHERS.

Additional color graphics may be available in the e-book version of this book.

Library of Congress Cataloging-in-Publication Data

ISBN: 979-8-88697-108-8

Published by Nova Science Publishers, Inc. † New York

> "... an enigma presents itself which in all ages has agitated inquiring minds. How can it be that mathematics, being after all a product of human thought which is independent of experience, is so admirably appropriate to the objects of reality? Is human reason, then, without experience, merely by taking thought, able to fathom the properties of real things? In my opinion the answer to this question is, briefly, this: as far as the propositions of mathematics refer to reality, they are not certain; and as far as they are certain, they do not refer to reality..."

– Albert Einstein, *Geometry and Experience (1921)*

> "Whatever is the lot of humankind I want to taste within my deepest self. I want to seize the highest and the lowest, to load its woe and bliss upon my breast, and thus expand my single self titanically and in the end go down with all the rest"

– Johann Wolfgang von Goethe, *Faust, First Part*

Contents

Acknowledgments   xiii

1   Au Lecteur: A Philosophical-Literary Journey through Time and Space   1
    1.1. A Crack in the Crystal Palace 2
    1.2. A Geopolitical Crisis 4
    1.3. A Short Excursus on Relativity in Literature 11
    1.4. The Enigma of Time in Art 15
    1.5. An Open Conclusion 18
    1.6. Book's Aim 19

I   **Special Relativity**   21

A — Towards Special Relativity   23

2   **Math Prerequisites**   25
    2.1. Something about Tensors 25
         Vector and their Invariant Character 25
         Vector as a Matrix 26
         Tensors 26
    2.2. Hyperbolic Functions 28

3   **The Crisis of Classical Physics: A Falling House of Cards**   29
    3.1. Aether Historical Theories in Brief 30
    3.2. The Michelson-Morley Experiment 30
         Arm Parallel to the Aether Wind 32
         Arm Perpendicular to the Aether Wind 32
         Conclusion 34

4 Definitions, Postulates and Principles — 35
- 4.1. Basic Definitions — 35
- 4.2. Galilean Transformation — 37
- 4.3. Newton's Fundamental Laws of Dynamics — 38
- 4.4. Basic Postulates and Principles — 39
- 4.5. Frame of References and Frame of Coordinates — 41

5 Where It All Began: The Light Clock — 47
- 5.1. Perpendicular Distances Do Not Change — 48
- 5.2. The Physical Apparatus — 49
- 5.3. Two Systems of References — 49
- Who Is Really Moving with Respect To? — 49
- 5.4. Train's Time — 49
- 5.5. Rail's Time — 50
- How Is It Possible to Measure It? — 50
- Straightness of Trajectory — 51
- 5.6. The Role Played by the Pythagorean Theorem — 51
- 5.7. Time Dilation? — 53
- 5.8. Distance Contractions — 54
- 5.9. Indirect Relativistic Measures — 55
- 5.10. Criticism to the Contraction of Distances — 55
- 5.11. Light's Climb Rate — 56
- 5.12. Aberration Angle — 57

6 The Lorentz Transformations — 59
- 6.1. Lorentz Space-Equation and Its Inverse — 60
- 6.2. Lorentz Time-Equation and Its Inverse — 61
- Finite Difference Lorentz Equations — 62
- 6.3. Properties of Lorentz Equations — 63
- Deriving Time Dilation and Space Contractions — 63
- Deriving the Composition of Velocities — 65
- 6.4. Invariant Interval — 66

7 Simultaneity and Causality — 69
- 7.1. Introduction — 69
- 7.2. Einstein's Train Paradox — 70
- Train Track's Reference Frame — 71

	Train's Reference Frame	71
	Time Interval Measured by the Train Track	76
	Time Interval Measured by the Train	77
7.3.	Train Paradox - Light Sensors' Variation	79
	Train Track's Reference Frame	79
	Train's Reference Frame	80
	Time Interval Measured by the Train Track	85
	Time Interval Measured by the Train	86
7.4.	Car and Garage Paradox	88
	The Nature of the Paradox	88
	Solving the Paradox	90
	Time Interval Measured by the Car	92
7.5.	Chronological Order	93
7.6.	Causality	96
7.7.	Conclusion	100

B — Special Relativity: Kinematics 101

8 Lorentz-Minkowsky's Spacetime 103

8.1.	Four-Position	104
8.2.	Four-Velocity	105
8.3.	Four-Acceleration	107
8.4.	Lorentz-Minkowsky's Metric	107
8.5.	Deriving with Respect to Proper Time	107
	Quoting Two Proper Quantities	108
	Quoting Two Non-Proper Quantities: the Celerity	109
	Deriving the Four-Position with Respect to Proper Time	109
8.6.	Universe Lines	110
8.7.	A Look towards Dynamics	111
	Energy	111
	Four-Momentum	112
	Four-Force	114
8.8.	Some Interesting Solved Exercises	115

9 The Accelerated Motion 135

9.1.	A Common Misconception about SR	136
9.2.	Definition of Uniformly Accelerated Motion	138

9.3.	Defining Four-Acceleration	138
	The Derivative of γ with respect to τ	138
	The Derivative of $\gamma\vec{v}$ with respect to τ	139
	Result	139
9.4.	The laws of uniformly accelerated motion	139
	The Square Norm of 4−Acceleration	139
	The 4−Acceleration in MITCF, namely w.r.t. τ, σ	141
	Attempts to Find out the 4−Acceleration w.r.t. τ, s	144
	Law of Motion and Its Worldline	145
	An Interesting Link with the SEP	153
	Relationships between Proper and Non-Proper Spaces	155
9.5.	Equation Summary	156
9.6.	Boundary Analysis	157
	When Speed Is Much Smaller than c	157
	When Speed Tends to c	157
9.7.	Still Parabolic Motion? No, Hyperbolic!	158
9.8.	Speed versus Time Graphs	161
9.9.	Rindler's Metric for L-M Spacetime	162
	Defining Rindler's Coordinates and Metric	162
	Metric Interpretation	164
	Time Dilation in Rindler's Metric	166
9.10.	Horizons	168
	Horizon's Analysis in Rindler's Coordinates	176
	Relative Relativistic Uniformly Accelerated Motions	178
9.11.	Moving Away in Opposite Directions	180
9.12.	A Little Excursion in Cosmology	182
	Universe's Shape	182
9.13.	Round Trip to the Edge of the Universe	184
	Photon Moving on a Rubber Carpet	185
	The "Hubble's Law" and the Expanding Universe	186
	Sky Will Be Forever Black	188

10 The Accelerated Twin Paradox — 191

10.1.	The Original Twin Paradox	191
	Explaining the Paradox	192
	An Everyday Paradox	194
10.2.	Context and Data	194

10.3. The Paradox No Longer Exists 195
      First Phase: Acceleration 198
      Second Phase: Uniform Rectilinear Motion 198
      Third Phase: Deceleration 199
      Solution 200

II   A Glimpse at General Relativity   203

11 Gravitational Lensing and Proofs of General Relativity   205
11.1. Geodesics 207
11.2. Tangent Spaces 210
      Christoffel Symbols Geometric Definition 211
      Christoffel Symbols Metric Expression 212
11.3. Schwarzschild Metric 212
11.4. Maximal Aging 214
      The Role of Acceleration 217
11.5. The Eddington Experiment 219
11.6. The Briatore-Leschiutta Experiment 221
      Solving the Problem: A Prediction of the Result 222
      This Paper-Sheet Is Too Wide! 223

III   Conclusion   225

12 Relativity in a Nutshell   227

IV   Appendices   229

A Hyperbolic Functions   231
A.1. Preamble 231
      Trigonometric Functions 231
      Hyperbolic Functions 231
A.2. Definitions 233
      Definition of $f(x) = \cosh(x)$ 233
      Definition of $f(x) = \sinh(x)$ 235
      Inverse Hyperbolic Functions 236

A.3. Defining the Hyperbolic Functions 236
A.4. The Hyperbolic Tangent 237
A.5. Full Geometric Interpretation 239

Bibliography   **243**

About the Authors   **245**

Index   **247**

Acknowledgments

We would like to thank Alessandro C. Ciano, Cosimo De Luca, Eva C. Salvatorini for the invaluable contributions they have made to the creation of this book. The book wouldn't have been possible without their help.

I, Raul, would like to dedicate this book to Eugenia, the only person I've ever loved and to thank all the Students for their incomparable help in understanding my being and the Universe around me.

We hope that this book will fascinate, amaze, dream ...

Chapter 1

Au Lecteur: A Philosophical-Literary Journey through Time and Space

This is not a simple introduction, with no purpose, only for aesthetics.
This goes deeper. And with fewer words.
Concise, short but substantial, as everything should be.
The goal of this premise is not to enlighten the minds of readers but to shed light on the philosophical, geopolitical, cultural effects of a physical theory that seems to have very little connection with reality.
Everything is extremely connected: Mathematics, Physics, Philosophy, Literature, and even History.
Aristotle with his theory determined the political hierarchy of a millennium.
Einstein in turn changed not only the vision of a society that was already shaking during the Bella Époque, but the political hierarchy of a world close to defeat. He brought to the fore new forces, irrational forces, which lay in souls waiting to be released.
Einstein is the spectre that roams Europe[1] of 1900, is the box of a domino

---
[1]Reference to the famous incipit of the Manifesto of the Communist Party: *"A specter is haunting Europe—the specter of Communism. All the powers of old Europe have entered*

waiting to be broken and crushed.
Einstein is the fourth wound to the narcissism of humanity[2].

1.1. A Crack in the Crystal Palace

Everything begins where everything ends.

On November 14, 1831, the gates of Hell opened wide in Europe: Darkness envelops everything and a chasm opens at every step.

Men, who had always had their eyes raised to heaven, lower their heads and discover for the first time the immense emptiness under their feet. Hegel is dead, systematic philosophy is dead, all certainty is dead.

This premise is essential to understand every event following the death of the philosopher from Stuttgart.
Hegel, in fact, more than any other philosopher, had tried to unite the multiplicity of reality in a single structural work that could explain everything:

> *"The truth is the whole. The whole, however, is merely the essential nature reaching its completeness through the process of its own development. Of the Absolute it must be said that it is essentially a result, that only at the end is it what it is in very truth; and just in that consists its nature, which is to be actual, subject, or self-becoming, self-development".*

His death, which took place in a period of great turmoil and revolutions, opens a new phase in the history of philosophy: Hegel carries with him the

---

into a holy alliance to exorcise this specter; Pope and Czar, Metternich and Guizot, French radicals and German police spies".

[2] Reference to Freud's essay *"A General Introduction to Psychoanalysis"* (delivered as lectures 1915–1917 and first translated into English in 1920), in which the author identifies the three great humiliations inflicted on humanity in the Copernican, Darwinian and psychoanalytic theory: *"Humanity has in the course of time had to endure from the hands of science two great outrages upon its naive self-love. [...] But man's craving for grandiosity is now suffering the third and most bitter blow from present-day psychological research which is endeavoring to prove to the 'ego' of each one of us that he is not even master in his own house, but that he must remain content with the veriest scraps of information about what is going on unconsciously in his own mind". [...]*

dream of man to grasp reality through universal ideals and values, and the void shows itself to man in its dramatic essence.

There is a philosopher in Gdansk who seems to have understood first of all the new vision of reality that was destined to rise in the following years. This philosopher is Arthur Schopenhauer and he identified the very principle on which the world depends in evil: he understood that "behind the wonders of creation" cannot hide supernatural entities or positive ideals, but rather *"there is an arena of tormented and anguished beings"*, that life is pain as an eternally unfulfilled desire and boredom as the impossibility of unlimited pleasure, that love is a mere stratagem of the genius of the species, that everything is but a *"Wille zum Leben"*, a will to live that overwhelms everything in order to perpetuate itself.

Schopenhauer is only the beginning, it is the first flame of a fire that will spread in every direction.
All philosophers will follow his path, burning up any residue of certainty that seemed to man to remain.

Feuerbach, taking up the Hegelian concept of *"alienation"*, demonstrates that religion is nothing more than an inverted anthropology, that it is man who creates God and not the other way around and that it is determined not by ideals but by matter itself: *"Man is what he eats"*.
Marx, in the wake of Feuerbach, overturns the idealist assumptions and demonstrates that it is the economic structure that determines the superstructure of society, and therefore also religion, showing how the exploitation of workers is actually hidden behind the regularly contracted labour relations.
Freud, starting from the certainties found through positivism, ends up denying the very concept of free will, through the discovery of the unconscious, which makes man *"aware that he is not even master in his own house"*.
Nietzsche strikes men in the heart with the destruction of the *"real world"*, announcing the death of God and the advent of a superman who aims at the *"transvaluation of all values"*.

And finally, a man arrives who, in 1905, five years after Nietzsche's delusional death, publishes a scientific article in the journal *Annalen der Physik* entitled *"On the electrodynamics of moving bodies"* (*Zur Elektrodynamik bewegter Körper*), where, in ten paragraphs, he expounds a theory that would later be known as *Special Relativity*.

That man is Albert Einstein, and he is the son of an age of uncertainty, the son of a shaky society, drowning in its own inability to make sense of itself.

By handing over the pages of the article, he does not realize that it will contribute to a sea change.

Einstein's is a real crack in the crystal palace, an earthquake that shakes the foundations and breaks the certainties with which the Bella Époque had deluded men.

He demonstrated that space and time are not absolute, that each observer finds himself in front of his own reality that does not coincide with that of the others.

Galileo collapses under Einstein's blows, Newton is overtaken, and everything loses sense.

There is nothing left for man to believe in.

1.2. A Geopolitical Crisis

It is 350 BC,

A philosopher from Stagira, Macedonia, rolls his eyes.

His soul merges with the cosmos, and the cosmos is revealed in his extraordinary mysteriousness.

The result of this mystical vision is the first, great, cosmological treatise in history, the one that in Latin will be translated as *"De Caelo"* (or *"De Coelo"* in medieval Latin).

The philosopher is Aristotle, and his theory will determine the centuries to come.

But it is a **wrong** theory, terribly wrong, which places the earth at the centre of a Universe created specifically for man, where every movement is given by the first substance, God: one and last pure act, thought of thought.

Au Lecteur: A Philosophical-Literary Journey ... 5

"... L'amor che move il sole e l'altre stelle" ("... The love that moves the sun and the other stars[3] ")

Leaving aside the characteristics of what will take the name of Aristotelian-Ptolemaic cosmological theory[4], it is necessary to dwell on the effects of the latter on a political, economic and social level.
Aristotle not only elaborated a cosmological model, a scientific theory, but he determined the very character of the society that would be born in the following years.
A hierarchical society, where social classes take the place of the Aristotelian spheres of heaven, eternally revolving around a God who is not in the hereafter, but in the hereafter, who is not unique, perfect and infinite but dramatically "human, too human[5]".

The certainties brought by Aristotle have allowed the creation of a top-down society, where the condition of the servant is rationally justified on the basis of a cosmological and otherworldly model that imposes and bends the population itself under its own weight. There is no redemption until after death[6], everything is inevitably and eternally equal to itself.

From these, albeit brief, considerations, it is easy to understand the very reason for the trial of Galileo brought by the Holy Church in 1633: The geocentric theory is fundamental to be able to maintain control of the social order, without it the hierarchy is broken.

No, no, no, no, no, no, stop, Galileo, stop!
For independent spirit spreads like foul diseases.
People must keep their place, some down and some on top!
Though it is nice for once to do just as one pleases.

---

[3]Dante, Paradiso, XXXIII, 14.
[4]Claudius Ptolemy, author of the treatise *The Almagest*, has taken up and perfected the Aristotelian cosmological theory, already studied by Hipparchus of Nicaea.
[5]Nietzsche, "Human, All Too Human. A book for free spirits", 1878 – 1879.
[6]The idea of death itself, understood as a passage towards bliss, is moreover exploited to maintain social order:
"Quando cade il soldin nella cassetta, l'anima vola in cielo benedetta..." (*"When the penny falls into the box, the soul flies to heaven blessed..."*).

Good people who have trouble here below
In serving cruel lords and gentle Jesus
Who bids you turn the other cheek just so
While they prepare to strike the second blow:
Obedience will never cure your woe
So each of you wake up and do just as he pleases!

Esteemed citizens, behold Galileo's phenomenal discovery:
The earth revolving around the sun![7]

And it is in this way that the destruction of the *"ordo ordinis"* carried out by Galileo, and even earlier by Copernicus, in turn, had many consequences on a political, economic and social level.

This premise is fundamental to understanding the effects of the Theory of Relativity on the society of the Bella Époque, which moves with the same dexterity of a blindfolded tightrope walker whose rope has been cut.
The following considerations want to present themselves as reflections on what is a phenomenon that arouses no little interest and wonder.

The starting point is once again an error:
The Theory of Relativity, in fact, should have been called *Theory of Invariants*. While destroying the Euclidean conception of previous space and time, Einstein does not base his theory on chance or on the impossibility of a scientific explanation of certain phenomena, far from it: *"God does not play dice with the Universe"* he will write in December of 1926 to the German physicist Max Born who was studying quantum mechanics. All Einstein's work tends towards the absolute.
In this regard, a translated text by Planck is reported, which seems to clarify the question.

Einstein's discovery that our concepts of space and time, which Newton and Kant placed at the basis of their thinking as absolute and given forms of our phenomenal intuition, have instead a relative meaning for the will that is implicit in choice

---
[7] The life of Galileo, Brecht, 1939.

of the reference system and method of measurement, is perhaps among those that most affect the roots of our physical thought. But denying the absolute character of space and time does not eliminate the absolute from the Universe, it simply moves it more back in the metric of four-dimensional multiplicity, which consists in merging space and time together in a unitary continuum by means of the speed of light. This metric is a thing in itself, detached from any will, and therefore is an absolute.

Thus also the theory of relativity, too often misinterpreted, not only does not suppress the absolute, but on the contrary highlights even more clearly that physics is always based on an absolute place in the external world. Since if the absolute, as many theorists of knowledge claim, existed only in the lived experience of each one, there should be as many physicists as there are physicists, and we could not at all understand why it has been possible, at least up to now, to construct a physical science. which is the same for the intelligences of all scientists, despite the differences in their lived experiences. It is not we who create the external world because it suits us, but it is the external world that imposes itself on us with elementary violence: this is a point on which it is necessary to insist, in our time steeped in positivism.

When, in the study of every natural phenomenon, we try to pass from what is particular, conventional and casual to what is general, objective and necessary, we do nothing but look for the independent behind the employee, behind the relative the absolute, behind the transitory the perennial. And, as far as I am aware, this trend is not only detectable in physics, but in every science, and not only in the field of knowledge, but also in that of good and beautiful.

I will conclude with a very obvious but embarrassing question. Who can guarantee us that a concept, to which we ascribe an absolute character today, will not prove relative tomorrow, and will not have to give way to a higher absolute concept? The answer can only be one: no one in the world can take on such a guarantee. Indeed, we can be sure that the absolute true will never be grasped. The absolute is an ideal destination that we always have in front of us without ever being able to reach it. This may perhaps be a thought that disturbs us, but to which we must adapt. Our condition is comparable to that of a mountaineer who does not know the mountains for which he walks and never knows if behind the peak he sees in front of him and wants to reach another higher one does not arise by chance. For him, as for us, it may be a consolation to know that we are proceeding further and further and higher and higher, and that there is no limit that prevents us from continuing to approach the goal. Pushing towards this goal ever further and ever closer is the true constant effort of every science, and we can say with Lessing that not the possession of truth, but the victorious struggle to conquer it makes the scientist happy; because every stop tires and ends up unnerving. A strong and healthy life only thrives on work and progress. From the relative to the absolute[8].

Physicist Richard Feynman himself mocked the *living room* philosophers who used to summarize the Theory of Relativity in two verbal formulas: *"Physical phenomena depend on the frame of reference"*, and *"Everything is relative"*. The first proposition is moreover banal and evident even before Einstein's intervention, while the second is, as already demonstrated, completely false and born from the wrong identification of relativity with relativism.

The geopolitical consequences of Einstein's theory, therefore, can all be derived from an error of assessment, from a misunderstanding.
On the other hand, it is likely to think that Aristotle himself, confronted with the consequences of his cosmological theory and that *"Ipse Dixit"* presented as absolute truth, would be horrified[9]. For this reason, it is nec-

---

[8]M. Planck, Knowledge of the physical world.
[9]Descartes, in his Discourse on the Method, writes of the Aristotelian's: *"They are like the ivy that does not tend to rise higher than the plants that support it and, even, often,*

essary to consider not so much Einstein's theory itself, but the perception that society had of it.

The society of the Bella Époque is a grandiose society of blind men who are heading with force towards the precipice.
It is the age of certainties, discoveries, innovations, peace, which hide behind them the horror of a reality that is close to being broken and destroyed. War is in the air but men are *"sleepwalkers"*, taking up Christopher Clark's definition from the book of the same name.

And it is precisely in the age of certainties that a feeling of instability and bewilderment is created, due to dialectical necessity.
Einstein takes the first big step by showing that space and time are not absolute, that space is not Euclidean, that all scientific knowledge of time is wrong.
It is the beginning of the end.

Italy is the country in which the Theory of Relativity has had the greatest influence.

In 1922, the writer and critic **Adriano Tilgher** published the booklet *Relativisti contemporanei (Contemporary relativists)*, celebrating Einstein as the *"leader of the formidable relativist assault which, radiating from Germany throughout the civilized world, tends to renew the very foundations of our knowledge"*, with the merit of *"having introduced subjectivism into the science of nature by means of physical-mathematical arguments"*, so that relativity was part of a broader movement of thought inspired by a *"Activist intuition of the world and of life"*, which in the political field found expression in fascism.
Mussolini's reply was not long in coming:

> *"Esattissimo! Con questa affermazione Tilgher immette il fascismo nel solco delle più grandi filosofie contemporanee: quelle della relatività".*

> (*"Exactly! With this statement Tilgher puts fascism in the*

when it reaches the top, it goes down again".

wake of the greatest contemporary philosophies: those of relativity")

Mussolini was evidently happy to give his movement an intellectual fashion label, proposing himself as the one who had created *political relativism*.
Even earlier, eleven months before the march on Rome, he had written:

> *"Se per relativismo deve intendersi il dispregio delle categorie fisse per gli uomini che si credono portatori di una verità obiettiva immortale, per gli statici [sic!] che si adagiano, invece che tormentarsi e rinnovellarsi incessantemente, per quanti si vantano di essere sempre uguali a se stessi, niente è più relativistico della mentalità e dell'attività fascista".*

(*"If by relativism is to be understood the contempt of fixed categories for men who believe themselves to be bearers of an objective immortal truth, for the static [sic!] Who recline, instead of tormenting and renewing themselves incessantly, for those who boast of always be equal to oneself, nothing is more relativistic than the fascist mentality and activity"*)
- Il Popolo d'Italia, 11/22/1921

He would then proceed the following year to the reprint of the contents that approached the philosophy of relative to fascism both in "Il Popolo d'Italia", with *Relativism and Fascism*, and in the first issue of his supplement "Gerarchia", with *Relativism and Politics*.
Mussolini would thus have contributed significantly to the diffusion of Einstein's theory in the political field.

Barbara J. Reeves in 1986 and 1987, he published two useful writings for the understanding of this phenomenon: *"L'appropriazione politica delle teorie della relatività di Einstein nell'Italia fascista, ovvero, come Mussolini può avere avuto un ruolo indiretto per lo sviluppo della fisica teorica in Italia"* (*"The political appropriation of Einstein's theories of relativity in fascist Italy, that is, how Mussolini may have had an indirect role in the development of theoretical physics in Italy"*) and *"Einstein*

Politicized: The Early Reception of Relativity in Italy".
The author documents the process that led to the spread of the theory of relativity in Italy, starting from its acceptance in academic circles, its acceptance in the Italian cultural and philosophical environment, up to its politicization.

The errors that led to this process were in the first place the language used by scientists in writings intended for the general public, in which the theory is defined with the terms of "revolutionary", "evolution", "destruction", "new construction", and subsequently the misidentification of relativity and relativism already presented.
Taking up the second error, it is possible to refer to the philosopher **Giovanni Gentile**, who, following the statements of Adriano Tilgher, expressed his appreciation for the theory of relativity, which in his eyes perfectly matched the relativism of contemporary reality. Relativity thus finds a direct correlation with phenomena such as Titanism, the revolt against tradition in the arts, imperialism and nationalist politics, and the First World War.

1.3. A Short Excursus on Relativity in Literature

If the Theory of Relativity has had consequences on a political and social level, it cannot fail to have strongly influenced the literature of the early twentieth century.
On the other hand, Galileo himself had contributed with his discoveries to the flourishing of overwhelming Baroque literature.

The starting point of this section is a reflection carried out by the philosopher, sociologist and critic of literature **György Lukács**, who, influenced by Hegelian philosophy, published in 1914 the essay on *"The Theory of the Novel"*.
He, defining in the first place philosophy as *"a sign of the substantial diversity of self and world, of the incongruity of soul and doing"*, analyzes the modern art form of the novel, which in his eyes is the very expression of splitting, separation, conflict.

"In modern times, the novel represents the orphaned condition

of the individual, who wanders around in a reality no longer illuminated by the light of the stars and who indeed suffers from being far away from any starry sky".

The novel is therefore the genre where more than any other man can reflect himself with his true nature as an orphan of a lost world, unable to make sense and make sense of himself.
But all literary genres cannot stand idly by watching the world collapsing in front of their eyes, they must take note of it and try to save themselves.

Einstein's scientific discoveries are but a further petty blow to a castle without foundations, which by falling overwhelms everything, and therefore, also literature.
These blows have thus contributed to deconstructing the figure of the intellectual, by now incapable of being the spokesperson of universal values and ideals: he is no longer able to believe even what his eyes see and his own actions.

Baudelaire had already expressed the *"loss of halo"* of the intellectual, but still attributed to him the possibility of knowing the mysterious *Correspondance* present in the *"forest of symbols"* of the world.

> *"What! You here, too, old pal? You, in this den of iniquity! You, quaffer of quintessences! You, who sup on ambrosia! This is a real surprise. My dear chap, you know my fear of horses and carriages. Just now, as I raced across the street, stomping in the mud to get through that chaos in motion where death gallops at you from all sides at once, my halo slipped off my head and onto the filthy ground. I'm afraid I didn't have the sang-froid to pick it up – let's just say I deemed it less disagreeable to lose my insignia than to have my bones broken. And then I said to myself, look for the silver lining. I can now walk around incognito, doing whatever nasty things I like, indulging my vices just as lesser mortals do. And here I am, just like you, as you see!*[10]*".*

For this reason the literary critic Romano Luperini writes:

---

[10]Loss of Halo (Perte d'Aureole), 1869.

"Everything is relative, everything becomes problematic there are no longer objective laws, laws depend on the observer's gaze. The observer's gaze decides on the norm which is the result of a researcher's mental process[11]".

One of the first to adapt his poetics to the new theory of relativity is **Luigi Pirandello**. Translating it into literature, he ends up declaring with extreme pain the destiny of the ego's inconsistency, which causes bewilderment and pain, which causes anguish. Life has now become *"a huge puppet"*.

"Whoever has understood the game, can no longer deceive himself; but whoever can no longer deceive himself, can neither enjoy nor enjoy life. So it is[12].

The arrival of the fourth dimension, that of time, also upsets the Trieste **Italo Svevo** who, in the *Conscienza di Zeno*, blends psychoanalysis and relativity in a magnificent novel where no information is reliable and is the reader himself which, through its own interpretation, tries in vain to construct a meaning where a meaning does not exist.
The time of consciousness, brilliantly defined later by **Bergson**, takes over the time of science, but even this is now relative: the protagonist of Svevo's novel moves in a deformed world, which is modelled under his feet, where time prevails over space.

"You see things less clearly when you open your eyes too wide"

This *"relativity of the world"* also occurs in **Federigo Tozzi**, in search of meaning where darkness reigns:

"When it is understood that religion is our soul, a profound part of our revelation, one less mystery, will return in honor of using Thomas Aquinas to more accurately interpret our soul.

---

[11] Romano Luperini, Didattica della Letteratura.
[12] Letter to his sister Lina, dated 31 October 1886 it presents the following main points: 1) Life is meaningless; 2) Writing and studying is similarly nonsense but serves as compensation for the frustration resulting from such discovery; 3) The ideals that help to live are self-deceptions or illusions. mystifying; 4) However, they are necessary to survive".

> We cannot deny the scientific discoveries, be they external or internal, indeed we must take them and give new nourishment to the fundamental complexity of our faith: the discoveries pass and go beyond, and are overcome; and our religiosity becomes more and more decisive and deeper, almost transported by voluptuousness intellectual of more and more renewing and modifying our knowledge. (...) It is not clear why the science of electric force should diminish our soul!"

But the meaning is not so easy to reach: the twentieth century seems to have definitively eliminated it from the life of man, in the exterior and interior.

The narrator of the twentieth century is then exemplified by **James Joyce** that *"he feels struck by facts that are insignificant in themselves, which are useless, and therefore they epiphanize, arrive at a manifesting power*[13]*"*.

In *Ulysses* time once again takes on a fundamental value: the subjectivity filter dominates unchallenged and is the true protagonist of the novel, through a *"stream of consciousness"* which imposes on the subject the task of creating the reality and no longer vice versa.

Virginia Wolf also elaborates her own personal vision of reality. In *To the Lighthouse* the author, starting from the cinematographic language and following in Joyce's footsteps, makes use of the two methods of the *"time montage"*, consisting in fixing the space by making the character's thought travel, and of the *"space montage"*, where instead it is time to be fixed and space to change.

Returning to Italy, **Giuseppe Ungaretti** demonstrates that he has understood, through his poems, a new conception of time and space that does not correspond to the Newtonian one: in *"I fiumi"* the author breaks the Aristotelian logic, replaced by a symmetrical logic where the principles of cause-effect, identity and the excluded third are no longer valid.

It is then the **Crepuscolari** and **Umberto Saba** who carry on the crisis of certainties that will be resumed in a refined and elegant way by **Eugenio Montale**, aware that the world of the modern is the world of *"splitting"*, constantly looking for a *"gap in the wall"* that he will never be able to find.

---

[13] G. Debenedetti, Il Romanzo del Novecento (The novel of the twentieth century).

Finally, a final author that needs to be mentioned is **Samuel Beckett** who in his play *Waiting for Godot* perfectly expresses the idea of a time that is no longer linear, but distressing and cyclical, which is no longer absolute but relative.
The sentence pronounced by the boy to the two vagabonds Vladimir and Estragon, *"Mr Godot will not come today, but will come tomorrow"* can only express the excruciating awareness of a world devoid of certainties, to which a wise reflection of a religious character.

> *"Let us do something, while we have the chance! It is not every day that we are needed. Not indeed that we personally are needed. Others would meet the case equally well, if not better. To all mankind they were addressed, those cries for help still ringing in our ears! But at this place, at this moment of time, all mankind is us, whether we like it or not. Let us make the most of it, before it is too late! Let us represent worthily for one the foul brood to which a cruel fate consigned us! What do you say? It is true that when with folded arms we weigh the pros and cons we are no less a credit to our species. The tiger bounds to the help of his congeners without the least reflexion, or else he slinks away into the depths of the thickets. But that is not the question. What are we doing here, that is the question. And we are blessed in this, that we happen to know the answer. Yes, in the immense confusion one thing alone is clear. We are waiting for Godot to come".*

1.4. The Enigma of Time in Art

The theory of relativity, for the reasons given in the previous sections, could only forcefully take over the Art.
On the other hand, art too has to face a new concept of reality, clashing with representations of the past. Space and time are relative and one cannot fail to acknowledge this revolution: once again the foundations are undermined and uncertainties and instabilities are brought about, leading to a new and wonderful vision of the world.

That the world was preparing to be shaken by the greatest earthquake

was however known to all even before 1905, and this idea is transmitted in the representations of the late twentieth century. The crisis of classical Newtonian physics is in the air, everywhere there is a feeling of change that does not go unnoticed:
Paul Cezanne challenges the conception of the picture as a simple and mere reproduction of reality.
Together with him are the **post-impressionist** painters, who, rejecting the naturalistic representation of reality, seek solidity in the image.
The flat and solid geometry is therefore found at the basis of their painting, binding with a level of use of colour.

In 1905, using a Nietzschean expression, *"God is dead"*:
It is the god of absolute spaces and times, which crumbles in the hands of a German scientist.
Art is deeply affected.

In fact, while Einstein elaborates his theory, it is **Pablo Picasso** who dominates the scene in the world of art. These are two events unrelated to each other but this does not mean that it is possible to speak of random coincidences: it is the same cultural climate, the same need for novelty, to go beyond the pillars of Hercules of empirical knowledge that pushes the two great minds.
Picasso destroys Euclidean geometry which has always been dominant in representations and, like Einstein, introduces the need to represent reality from a fourth point of view: that of time. **Cubism**, in fact, breaking the conventions on the uniqueness of the point of view, produces an effect similar to that brought by Einstein in the artistic field.

However, time in Cubism is a slow time, where everything is static.
But the static is not loved by everyone:
Already **Marchel Duchump**, in *"Nude descending the stairs"*, brings to the fore a new conception of time, representing no longer still bodies but the same subject broken down into several points of view and repeated in successive moments.

> *"Painting shouldn't be just retinal or visual; it should have to do with the gray matter of our intellect, instead of being purely visual".*

Finally, here is **Futurism**.

"The magnificence of the world has been enriched with a new beauty: the beauty of speed".
"To reach the futurist conception of the provisional, of the fast and heroic continuous effort, it is necessary to burn the black cassock, a symbol of slowness, and to melt all the bells to make them the rails of new ultra-fast trains".

This is how he writes **Filippo Tommaso Marinetti**, in the *Manifesto of Futurism*, published in French in Paris in the newspaper Le Figaro with the title of *Manifeste du Futurisme*.
We, therefore, have the cult of time, of a fast time, of a dynamism that pervades everything and deforms the image of things.

But the theory of relativity also contributes to the birth of **Metaphysical Painting**, of which **Giorgio de Chirico** represents one of the absolute masters.
We find ourselves in front of representations devoid of any temporal reference, where everything is immobile, where reality, time, is an enigma.

In the *"Enigma dell'ora"* (*"The Enigma of the Hour"*) of 1911 it is possible to notice how the entire space of the canvas is occupied by an immense arcaded portico, overlooking a large square, in which a motionless man, illuminated by the sun's rays, waits, enigmatically, next to a tub that opens into the ground, a symbol of death.
Time is the protagonist of the scene: the clock marks the 14 : 55 but the shadows serve to contradict the time represented. This contradiction is but the contradiction itself between the time of science and the time of life, already wonderfully described by Bergson.
De Chirico, therefore, focuses on the philosophical-existential condition of the eternal present, of empty, suspended time, terribly similar to Nietzsche's concept of Eternal Return.

However, the artistic movements presented so far never make their link with the Theory of Relativity as explicit as does **Surrealism**:
The melting clocks of **Salvador Dalì** are but an artistic representation of the latter.

Time dissolves and ceases to be conceived as a mere linear succession of instants. Therefore, even the space is deformed, preventing a univocal vision of the two quantities.
Past, present, future merge together indissolubly, human memory continues to persist as an accumulation of dream contents.

> *"Time is the delusional and surrealist dimension par excellence"*

Dalì states in an interview in the magazine *Minotaure* in 1935.

Another noteworthy author is certainly **Maurits Cornelis Escher**, who in his works strongly brings the themes of relativity and the applications of a geometry that, far from wanting to be Euclidean, becomes hyperbolic.

He is often traced back to **Optical Art**, as the optical illusion represents one of the main objects of his paintings.
Relativity, dated 1953, represents an image made up of many possible scenes, as many as the eyes of the observers.

1.5. An Open Conclusion

The goal of this premise is not to reveal anything but to make people reflect:
reflect on the consequences of a butterfly's wing flapping.
Surely the young Serbian student *Gavrilo Princip* while starting from the magazine of his pistol the two shots that would have killed the archduke Ferdinand and his wife did not have in mind Einstein and the theory of Relativity, of which he had probably only vaguely heard about. But his actions can be traced back to that larger period of uncertainty in which Einstein was formed and which Einstein himself contributes to bringing down, through another, furious blow to the foundations of an already crumbling castle.

Marx had enlightened minds by describing the dependence of the superstructure on the economic structure: here is asked to make a greater effort, to overturn Marx's assumption and to consider the Theory of Relativity as a crack that quickly spreads into a nearby Pandora's box about to open, like the forbidden apple that gives knowledge but condemns man forever.

With this premise the book begins, with this premise it is necessary to look at life:
Never reflect from a single point of view, from a single angle, but always look at the bigger picture.

1.6. Book's Aim

When a book can be defined as original?

Writing represents what distinguishes each man from the other, the unique and unrepeatable act that distances him from the beast and brings him closer to the divine.
Is it possible to apply the same to a physics book?
If so, then when a physics book can be defined as original?
The question arises: *are these pages really worth reading?*

The answer to this question is more complex than one might imagine.
This book tries to expose a theory known to everyone by changing point of view, renewing and simplifying concepts, known but not always understood.
Sometimes it's just a matter of perspective.
This book aims to describe in simple words some aspects of the theory that contributed to annihilating past certainties and building a new cognitive system.

Nature is eternal, the laws of physics will never be.

In this respect, is fundamental to remember the principle of falsifiability introduced by the Austrian philosopher and epistemologist Karl Popper, one of Einstein's most respected authors:

"*There can be no ultimate statements science: there can be no statements in science which can not be tested, and therefore none which cannot in principle be refuted, by falsifying some of the conclusions which can be deduced from them*[14].

A firm warning to all those who decide to undertake any course of study should be the famous Einstein's phrase:

"Everything should be made as simple as possible, but no simpler[15]*"*.

And this is the principle to which this book aspires.

The book is intended for all those who are interested in deepening the issues of Relativity by going outside the box in an intuitive and never banal way.

It is a book intended for everyone.

To those who have always wanted to understand Relativity but have never been able, to those who have never managed to disentangle themselves between the series of only apparently paradoxical arguments it proposes, to those who want to deepen their knowledge, to those who are curious, to those who never give up.

Dedicated to those who love knowledge above all.

---

[14] The Logic of Scientific Discovery (1959), 47.
[15] Attributed to Albert Einstein by Roger Sessions in the New York Times in 1950.

Part I
Special Relativity

Subpart A
Towards Special Relativity

Chapter 2

Math Prerequisites

This short chapter aims to illustrate the main mathematical prerequisites necessary to continue reading the book. The concepts of vectors and tensors, fundamental in the world of physics, will be analyzed.

Surely Mathematics is not a science, but an art form, useful in and of itself. We human beings do not know if it is immanent in the Universe or completely terrestrial. Anyway, nowadays Mathematics' structures seem to be particularly appropriate to penetrate the inner structure of the Universe.

Keywords: Vector, Matrix, Versor, Pinor, Tensor, Scalar, Hyperbolic functions

2.1. Something about Tensors

Vector and their Invariant Character

Any element of a vectorial space, which is a set satisfying certain axioms, is said to be, by definition, a **vector**. According to this definition - *a vector is all and nothing simultaneously* - in the sense that this definition does not catch at all the intimate nature of a vector, which is strongly connected with the notion of *invariant*. Assuming the concept of *coordinate system*, which is a *base* of the vector space, say V, a vector is an element in V such that, for any assigned *two* systems of coordinates, the linear combinations with certain coordinates of the first system *coincides* with the linear combination with other coordinates of the second one.

Vector as a Matrix

According to an analogue definition of vector, which goes in the direction of the modern mathematics - more oriented to functions and their applications, for any fixed base (e_i) in V, a vector $v = v^i e_i \cong (v^i)$ in V (here assumed to be real) is a $1-$linear function from V to $\mathbb{R}$ such that, for any $w = w^i e_i$ in V,

$$v(w) = \begin{pmatrix} v^1, & \ldots, & v^n \end{pmatrix} \begin{pmatrix} w^1 \\ \vdots \\ w^n \end{pmatrix} = \sum_{i=1}^{n} v^i w^i \in \mathbb{R}.$$

In other words, after having fixed a base of V, the vector v is here seen as a row, ready to be row-column multiplied (applied) for (to) any other $n-$uple of coordinates (or vector, module the above shaded isomorphism).

For any given vector space V endowed with a scalar product, which is a symmetric, bilinear form "b", a **versor** is a non null vector, namely a vector such that $b(v) \neq 0$. So, a versor, also said to be a *non-isotropic* vector, is a vector that provide a direction in the space V.

A **pinor** is defined to be a particular versor, which is a versor whose length (with respect to the above mentioned bilinear form b) is equal to one.

Going on with the definition of vector it could also be defined the notion of *spinor* and also of *rotor* and *twistor*, which, on the other hand, will not be useful in the sequel, so does not constitute a truly math prerequisite.

Tensors

Tensors are the obvious generalizations of vectors to multiple dimensions. As seen above, after having fixed a base in V, vectors (or $1-$tensors) can be identified with a list of numbers (ordered in a row) in the above described sense. Two-tensors are instead plus or minus naturally identified with a particular matrix of number, satisfying an analogous property with respect to vectors, 3−tensors are cuboids of numbers and so on.

More precisely, a 2−tensor is a function that is linear in all its two arguments. The number of arguments, here *two*, is said to be the *rank* of the tensor. Each of the two arguments can belong to a different vector space. For example, for any two given vectors $u = u^i e_i \cong (u^i) \in U$ (n−dimensional) and $v = v^i e_i \cong (v^i) \in V$ (m−dimensional) the 2-linear (or simply *bilinear*) function denoted by $u \otimes v$ can be isomorphically identified (as already done for $v \cong (v^i)$) with the *Kronecker product* between matrices u and v:

$$u \otimes v \cong \begin{pmatrix} u^1(v^i)^t & \cdots & u^n(v^i)^t \end{pmatrix} = \begin{pmatrix} u^1 v^1 & \cdots & u^n v^1 \\ \vdots & \ddots & \vdots \\ u^1 v^m & \cdots & u^n v^m \end{pmatrix} \quad (2.1)$$

so that for any $a = a^i e_i \in U$, $b = b^i e_i \in V$

$$(u \otimes v)(a,b) = \begin{pmatrix} b^1 & \cdots & b^m \end{pmatrix} \begin{pmatrix} u^1 v^1 & \cdots & u^n v^1 \\ \vdots & \ddots & \vdots \\ u^1 v^m & \cdots & u^n v^m \end{pmatrix} \begin{pmatrix} a^1 \\ \vdots \\ a^n \end{pmatrix}. \quad (2.2)$$

Conversely, since the spaces $\text{Hom}(U,V)$, $U^* \otimes V$ and so $U \otimes V$ are isomorphic, there is, in particular, a (linear) bijection between the spaces of rectangular matrices and $U \otimes V$, so that any given matrix can be viewed as the Kronecker product between two vectors. In this sense a matrix is a tensor and so is the associated matrix to a bilinear function. This is well known to be multiform depending on the choices of the basis of the factor spaces, on the other hand, the result turning from the multiplication in (2.2) does not change: *this is the nature of invariance characterizing the tensors in action*. Equivalently, in order to see it in another way, standing the above mentioned bijection, since for every matrix they can be found vectors u and v as in (2.1), since u and v are invariant as stated in § 2.1., a bilinear matrix and so a 2−rank tensor has an invariant character.

What about the associated matrix to a 3−linear function? Pushing on the above argumentation the Kronecker product between three vectors $u \otimes v \otimes w$ can be regarded as the following 3−dimensional matrix:

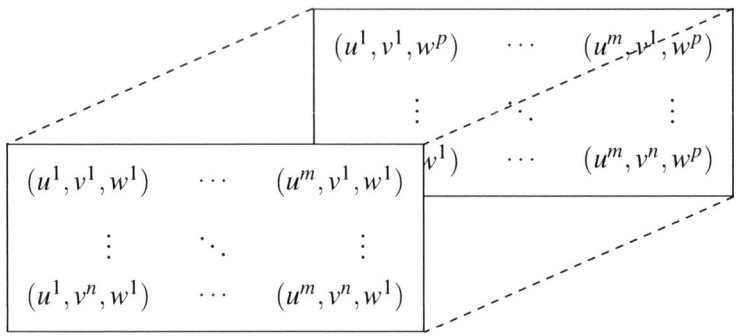

so that $(u \otimes v \otimes w)(a,b,c)$ consists in the iteration p−times of the product (2.2), where p coincides with the dimension of the space W which w belongs to, all these iterations gives a vector of p numbers, which ordinary multiplied by the vector c gives the result scalar in $\mathbb{R}$.

At last, in order to close the circle, what about **scalars**? They are merely zero-tensors, namely *numbers* which are *invariant* with respect to two any particular choices of system of 1−coordinates.

2.2. Hyperbolic Functions

In order to understand without difficulty the chapter related to accelerated relativistic motion, namely § 9, it is necessary to have a good knowledge of hyperbolic functions. Given the importance of the topic, please refer to the appendix located at the end of the book, § A.

Chapter 3

The Crisis of Classical Physics: A Falling House of Cards

This chapter presents the experiment that could be considered the precursor to the Theory of Relativity: "the Michelson-Morley experiment", which led to the proof of the non-existence of the aether. It is analyzed and described, placing importance on its consequences.

Keywords: Aether, Interference, Michelson-Morley experiment

One step away from the abyss. This is the condition of classical physics at the end of the twentieth century: a small blow is enough to make it collapse on itself.

Below the main aspects of Michelson-Morley experiment, which is the first blow to classical physics, will be deepened. It will be followed by the one inflicted by Einstein, which - as seen in the premise - will contribute to spreading uncertainty and anguish not only in the field of physics but in all fields of knowledge. Michelson and Morley refused to accept dogmatically the reality proposed to them and decided to investigate nature in search of answers, which will not be long in coming.

"The interpretation of these results is that there is no displacement. The result of the hypothesis of a stationary ether is thus shown to be incorrect, and the necessary conclusion follows that the hypothesis is erroneous".
- Michelson, 1881

3.1. Aether Historical Theories in Brief

According to historical testimonies the ancient Greeks of Plato's Academy were the first to express the concept of *aether*. *Plato* mentions it in the Fedone: *"The actual Earth, the pure Earth hovers in clear sky, were stars are, in that part called aether".*

It is possible to find a larger development with *Aristotle*, who made a systematic treatment of it, claiming that the aether was completing the other four elements, connecting the human microcosm with the universal macrocosm: *"An extremely tenuous substance, present in every part of the Universe, eternal, immutable and weightless, the aether was filling every void".*

This concept will be carried out by philosophers, writers and scientists for all the Middle-Age and the Renaissance, supported by the alchemical tradition. Experiments were attempted in order to isolate the element, thinking that it had magical properties, such as the capacity to transform the metals in gold and to protect from diseases. In 1730, during an experiment led with sulfuric acid and alcohol, the German physicist and alchemist *A. S. Frobenius* observed the quick evaporation of that volatile liquid that was associated with the "Spirit" turning to the sky, to the Aristotelian *cosmic ether*; he called that compound "Spiritus Aethereus": nowadays, the ethyl alcohol is still called *aether*.

Only by the beginning of the 17th century, *Descartes* discussed the concept with a more modern theory, asserting that the fifth element was the conduit for the light, attributing it the phenomenon of gravity also.

3.2. The Michelson-Morley Experiment

In 1888 *Albert A. Michelson* and *Edward Morley* intended to demonstrate the existence of the aether, arguing that, if the Earth travels around the Sun suspended in the aether, it must be subject during its movement to a *wind*, in the same way as those who run moving the air.

According to this theory, the speed of light depends on the motion of the Earth so that, if the light uses the aether to propagate then - in order to obtain two different results proving the existence of the aether - it is sufficient to compare the following two speeds of light:

(i) when it travels in the same direction as the Earth, namely, in favour of the aether wind

(ii) when it travels perpendicular to the aether wind.

The two scientists, therefore, built a system of doubling a light beam, whereby half of it continued in the same direction in which the Earth moves, the other half advanced in a direction perpendicular to Earth motion: the two rays would be affected by the *wind of aether* in a different way, so that, while travelling at the same speed in the beginning, they would not have had the same speed upon the arrival on the screen.

The instrument they used was the interferometer designed by Michelson himself in 1881 and improved in 1888 together with his colleague Morley. The interferometer is aimed to separate a single light source into two rays, which, after having travelled through two different optical paths, recombine forming an interference pattern, which can be observed on a screen. In detail, the light is transmitted to a semi-reflective mirror, also called *beam splitter*, splitting it into two parts. The first ray, reflected at an angle of 90° with respect to the incoming direction of incidence, proceeds towards a plane mirror B which reflects it backwards; the second ray, after having passed through the semi-reflective mirror, is reflected by another plane mirror C towards the same mirror that divided it. The superposition of the two rays gives rise to the phenomenon of interference. A converging lens, placed between the laser source and the beam splitter, ensures that the interference pattern has such a dimension to make the fringes, and their supposed movement, clearly distinguishable.

What is the reason why interference fringes are generated? According to Figure 3.1 the light source, being a laser one, satisfies the condition of *coherence*, so, if the lengths of the two arms L_1 and L_2 were the same no interference pattern would be formed.

By rotating the system, the fringes formed by the interference should have varied, as their inclination was different from the presumed aether wind. So, the entire system placed on a sandstone base that floated in a large tank filled with mercury. The system could thus rotate freely without any relevant friction, so as to remain in motion for very long times in order to give the possibility to perform a large amount of measurements.

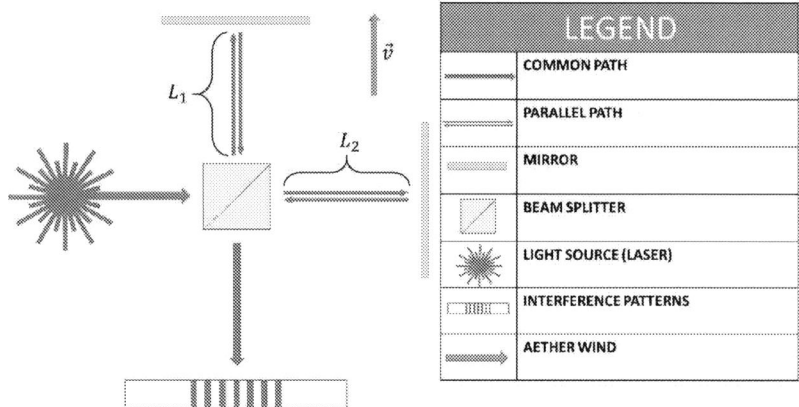

Figure 3.1.

Arm Parallel to the Aether Wind

The forward speed of light along the segment from A to B is equivalent to $c+v$, being in favour of the aether wind. On the return, along the segment from B to A, travelling in the opposite direction to the previously mentioned wind, the speed of the light ray should be instead $c-v$.

Standing the above, the time taken to travel the same length L_1 with two distinct speeds is as follows:

$$\Delta t_1 = \frac{L_1}{c+v} + \frac{L_1}{c-v} = \frac{2L_1}{\frac{c^2-v^2}{c}}. \tag{3.1}$$

Arm Perpendicular to the Aether Wind

Regarding the arm perpendicular to the aether wind (whose length, remember, is L_2) it affects the motion of light differently. In detail, from outside

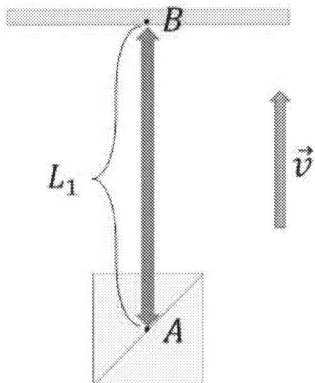

Figure 3.2.

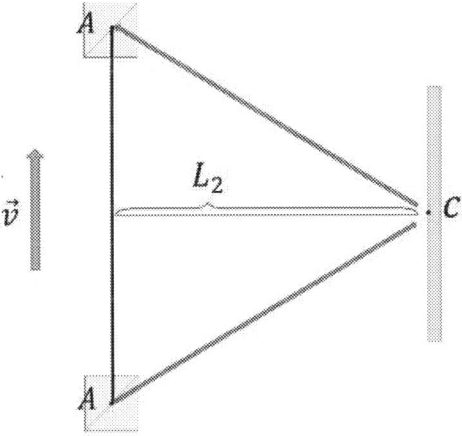

Figure 3.3.

the Earth, staying fixed together with the aether, the experimental apparatus appears in motion (in this specific case - *upwards*) moving together with planet Earth at speed $\vec{v}$ with respect to the aether (which, remember, is here thought to be fixed). The experimental apparatus blown by the aether wind is exemplified in Figure 3.1. In Figure 3.3 the beam splitter A is represented twice (in the instants the ray of light leaves and arrives on it), the mirror being represented only once in the (middle) instant the ray of light hits it. The light velocity vector is represented *oblique*, as a matter of fact it is the sum of the vectors $\vec{v}$ (upward) and $\vec{c}$ (horizontally directed). In other words light travels along two oblique trajectories, although it starts and returns to the same point A, being the system itself translated upward.

In order to compute the space travelled by light it will be used the Pythagorean theorem. By indicating with t the time needed by light to reach the point C it results:

$$\overline{AC}^2 = (ct)^2 = (vt)^2 + L_2^2 \quad \Leftrightarrow \quad L_2^2 = c^2t^2 - v^2t^2 = t^2\left(c^2 - v^2\right)$$

if and only if

$$t = \frac{L_2}{\sqrt{c^2 - v^2}}$$

where $\sqrt{c^2 - v^2}$ is the speed of light along each of the two oblique paths.

Standing the above, the time taken by light to travel this route, namely to reach point C and returning to A is:

$$\Delta t_2 = \frac{2L_2}{\sqrt{c^2 - v^2}}. \tag{3.2}$$

Conclusion

In conclusion, it can be deduced that - with the presence of the aether wind - the times taken by the two light rays (3.1) and (3.2) are different from each other.

However, what is decisive among the results obtained is the lack of variation of the interference fringes. If the aether existed, the interference fringes would not be static with respect to the rotating system. As the inclination varies, in fact, there would necessarily be a change in the action of the aether wind on the two light beams.

This experiment was surely an important anticipation of Einstein's theorization of the invariability of the speed of light in vacuum.

Chapter 4

Definitions, Postulates and Principles

This chapter is intended to give a sort of structure to the other ones. It is therefore intrinsically rich in references to the entire book in which the definitions, postulates and principles will be gradually deepened.

Keywords: Law, Principle, Axiom, Postulate, Frame of coordinates, - of reference

4.1. Basic Definitions

Definition 4.1. *In Physics, a **law** is a statement that comes from observations about Nature. If the statement starts out theoretically then the tendency is to refer it as a **principle**.*

Some examples of principles are the well known *three principles of thermodynamics, Archimedes' principle* and so on.

Definition 4.2. *In Mathematics and Science, such as Physics, an **axiom** is a truth assumed to be true without any need of any proof. A **postulate** is an axiom with a strong intrinsic geometric significance.*

Definition 4.3. An *event* is something that take place in a well defined place at a well specified time.

For more details see also § 8.1.

Definition 4.4. A *Frame of Coordinates*, also said a system of-, is a mathematical aid particularly useful in order to describe physical phenomena. Some examples of them are the Cartesian coordinates, polar, cylindrical and spherical coordinates and so on any coordinates of general relativity, see further.

Definition 4.5. A *Frame of Reference*, also said a system of-, is a laboratory, here intended as a site where physical measures can be performed. It is made of one or more bodies whose reciprocal distances are invariant. Some examples are the following: a room, the human wrist (where time measures can be performed, by taking into account the blood pulse, for example) a train wagon and more in general every rigid body.

Definition 4.6. The *proper length* between two events is the length measured in the frame of reference in which the starting and ending positions (and all the positions between them) are determined at the same time.

So, if someone is running in a valley and sees a leopard still on a tree he cannot determine the proper distance between him and the leopard. As a matter of fact, it is not possible to determine *at the same time* the position of *all* the points that separate him from the feline.

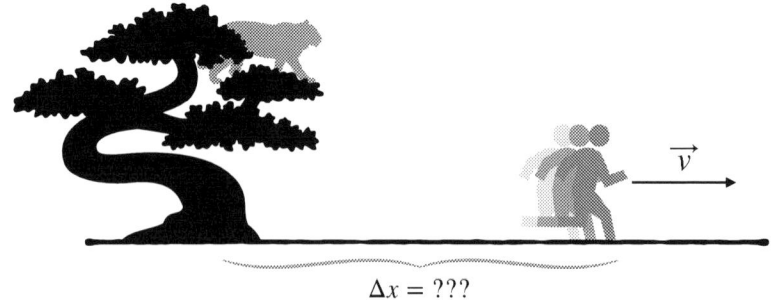

$\Delta x = ???$

Figure 4.1.

Definition 4.7. *The **interval of proper time** between two events is the time interval measured in the frame of reference in which the starting and ending instants (and all the instants between them) are determined in the same position.*

In other words, to define the *interval of proper time*, the notion of space is needed, *the same position*; symmetrically, to define the notion of *proper length*, the notion of time, *the same instant*, is needed. Time and space are so intrinsically linked. Note that both the above definitions are strictly referred to a particular frame of reference: if the frame of reference changes then the interval of time with respect to it is referred needs no longer to be proper. The same conclusion regards the notion of *proper length*.

In order to appreciate the contents presented within the various chapters, it is necessary to start climbing on the giants' shoulders to gain an adequate perspective without which it is not possible to build any new theory. Some of the essential aspects for understanding are presented below.

4.2. Galilean Transformation

First of all, to understand the earthquake caused by the birth of the Theory of Relativity, it is useful and interesting to remember the laws that were believed to be certain and that in some way were swept away by the shock wave caused by Einstein. These are obviously Galileo's transformations.

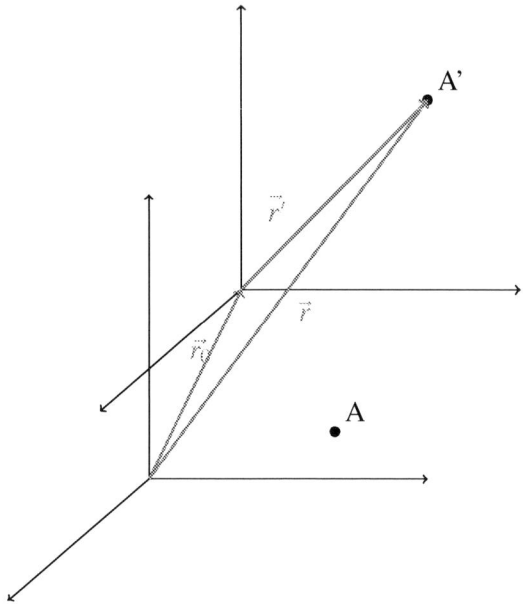

Given two frame of reference, one fixed (x,y,z,t) and the other (x',y',z',t') in motion with respect to the first, referring to the above Figure, standing the fact that $\vec{r}' = \vec{r} - \vec{r}_0$, the coordinates of the second frame with respect to the first one can be expressed as follows:

$$\begin{cases} x' = x - vt \\ y' = y \\ z' = z \\ t' = t. \end{cases} \quad (4.1)$$

It is necessary to pay attention to the last equation, namely $t' = t$, which still places time as an invariant between different reference systems.

4.3. Newton's Fundamental Laws of Dynamics

Here are reported Newton's fundamental law in their original version, both in Latin and in English.

- **Lex I/First Law.** *Corpus omne perseverare in statu suo quiescendi vel movendi uniformiter in directum, nisi quatenus a viribus impressis cogitur statum illum mutare.*

A body at rest persists in its state of rest, and a body in motion remains in constant motion along a straight line unless acted upon by an external force.

- **Lex II/Second Law.** *Mutationem motus proportionalem esse vi motrici impressae, et fieri secundum lineam rectam qua vis illa imprimitur.*

The alteration of motion is ever proportional to the motive force impressed; and is made in the direction of the right line in which that force is impressed.

This law can be summarized in the following equation, which is of fundamental importance:

$$\vec{F} = m\vec{a}. \qquad (4.2)$$

- **Lex III/Third Law.** *Actioni contrariam semper et æqualem esse reactionem: sive corporum duorum actiones in se mutuo semper esse æquales et in partes contrarias dirigi.*

To every action there is always opposed an equal reaction; or the mutual actions of two bodies upon each other are always equal, and directed to contrary parts.

4.4. Basic Postulates and Principles

Principle 4.8 (Correspondence Principle – CP). *The behaviour of physical systems described by the Theory of Relativity coincides with the analogous one described by Newtonian classical physics in the limit of low velocities and small masses.*

Postulate 4.9. *(First of Special Relativity) The laws of physics are the same in any inertial frame of reference.*

Remember that, in Special Relativity (SR), the *only* admissible frames of reference are *all* the inertial frames. In General Relativity (GR) instead, the admissible frames of reference are *all* reference frames.

Postulate 4.10 (**Second of Special Relativity**). *The speed of light, in empty space, is constant in all inertial frames of reference. Its constant value is exactly c equal to* $299\,792\,458\,\text{m}/\text{s}$.

Regarding this postulate see also § 6.3.

Problem 1 (Superluminal speed - colloquial treatment). *Sometimes, in physical literature it seems that the maximality of the speed is a peculiarity of light in vacuum, with constant value c equals to* $299\,792\,458\,\text{m}/\text{s}$, *with respect to any inertial frames. Is it equally true in accelerated systems? In other words, in the maximum expression of Relativity - the so-called General Theory of Relativity - is it possible in the Universe to reach a speed higher than the speed of light?*

Solution 1. Substantially neither Special nor General Relativity permit to walk faster than $299\,792\,458\,\text{m}/\text{s}$. Exceptions are some phenomena predicted by General Relativity and Quantum Mechanics, such as:

- inflation, referred to the process of spacetime expanding according to the law expressed by equation 9.54, see further;

- Einstein-Rosen bridges, also known as *wormholes*, and *warp drive*, which, however, requires lots of ordinary matter (aimed to shrink the front) and exotic matter aimed to expand the rear side of the spaceship, which is, by now, completely unknown;

- entanglement. □

Definition 4.11 (**second**). *Specially magnetized nucleus*[1] *such as cesium ones (in detail* $^{133}_{78}\text{Cs}$ *isotope) interact with the magnetic field generated by their electron clouds (exactly with the only electron belonging to the* $6s$ *orbital) so that they tick with a specific frequency.*

In detail, by analyzing the frequency emission spectrum ($f = \Delta E/h$ *) of the* $6s$ *electron from the hyperfine level* $HF = 4$, $F = 0$ *to* $HF = 3$, $F = 0$ *one*[2], *it can be detected such an hyperfine transition-tick of the electron. By counting* $9\,192\,631\,770$ *of these ticks the elapsed time is called one second.*

---

[1] Such a magnetization is essentially due to protons and neutrons' spin.
[2] Such effect of splitting of an *emission* spectral line into different components is called *direct Zeeman effect*.

Definitions, Postulates and Principles 41

Definition 4.12. *The* **metre** *is the length of the path travelled by light in a vacuum in* $1/299\,792\,458$ *s.*

Definition 4.13. *The* **light-second** *is the length of the path travelled by light in a vacuum in* 1 *s.*

Definition 4.14. *The* **light-meter** *is the time necessary to the light to travel the space in vacuum of* 1 m.

4.5. Frame of References and Frame of Coordinates

Spacetime reference changes play an analogous role with respect to coordinate changes.

Suppose to be in a certain inertial frame of reference K where the electric field $\vec{E}$ is identically equal to zero; on the other hand, suppose the existence of a magnetic field $\vec{B}$. Let's consider a charge q and let it move at constant velocity $\vec{v}$: for example a ball made of elder, stuck in a rectilinear wire and having uniform velocity. The magnetic field acts on the ball by exerting the Lorentz force given by the well-known expression $q\vec{v} \times \vec{B}$, which is balanced by the constraint's reaction (exercised by the wire). Now, let's analyze the physics in the frame of reference K' in which the ball is stationary: here the Lorentz force vanishes since $v \equiv 0$, on the other hand, since the constraint's reaction is actually present (and continue to wear out the wire), since the ball is stationary, there must be another force acting on the ball. Since the ball is charged, this other force must be an electrical one. So, the presence of only a magnetic field in K, in which the ball is moving, imply the presence of also an electric field in K', in which the ball is stationary. *Conversely*, if in K there is only an electric field generated by a certain number of fixed charges, then in K' these charges are moving, so that they generate a magnetic field.

Consider now on planet Earth a frame of coordinates with the z axis oriented upward: then there exists a gravity field directed along the z axis, but there is no gravity field along the two other axes, x and y. Let's now change the frame of coordinates by rotating the previous one around the y axis of a certain angle: in this new frame of coordinates, the gravity field exists both along the z and the x axis. This is absolutely obvious, but less obvious if compared with the previous discussion: the modification of the coordinate frames is covered here, while previously the modification of the reference frames was treated. The conclusion is that changing in reference play an analogous role with respect to coordinate changes. And this is true not only in the ordinary 3−Euclidean space but also in a curved spacetime. Besides, the analogy between the two above prospected situations goes further: the electric field and the magnetic one are not two separate entities but two aspects of the same physics entity, namely the *electromagnetic field*, just like g_x, g_y and g_z are aspects of the same entity: the g field. As a matter of fact, SR and so GR was born starting from the analysis of the Lorentz force.

Definition 4.15. *An **inertial frame** is a frame of reference in which the net external force, which is, the sum of all the external forces acting on it, equals to zero. So, invoking Lex I and the definition of relativistic momentum, see § 8.7., an inertial frame is a frame in uniform rectilinear motion. As a matter of fact, $d(\gamma mv)/dt = 0$ implies $\gamma mv = b$, where b is a suitable constant, so, without losing in generality it results $\gamma v = b$. Now $x \mapsto x \cdot (1-x)^{-1/2}$ is an injective map, so, for any fixed b, v is uniquely determined. It follows that, by generalizing the argumentation to the three ordinary dimensions for speed, $\vec{v}$ is uniform, namely the motion is uniform in direction.*

An ask arises naturally: the above mentioned *uniform rectilinear motion* refers to which frame? It does not refer to a particular one but to all the frames in relative uniform motion with respect to each other with the same speed. So, being this a relation of equivalence, there is exactly one equivalent class for each speed.

Principle 4.16 (**Special Relativity Principle**). *Inside a Galileo's ship[3] cruising with uniform motion along the same direction, no one can distinguish whether or not he is carried along in such a motion.*

Origin of the name *Special Relativity (SR)*. The influence of such a principle in Special Relativity is so strong that Max Planck named the theory after the principle. Einstein, however, should have preferred to stress the invariance aspect of the theory. The transformations between two

---

[3]Which is a ship without upper deck, portholes or communications with the external world.

frames of reference of SR are the *Lorentz transformations*, which consists in transformations between *two inertial frames of reference - two frames of coordinates* moving one respect to each other in uniform motion. The theory regarding Lorentz transformation has been developed starting from § 6.1..

Principle 4.17 (Special Covariance). *Under the transformations between two frames of reference of SR the equations of physics remain the same.*

The satisfaction of the principle *of special relativity*, namely the principle of *special covariance*, is established by often tedious algebraic manipulation: the equations of the theory are transformed under the Lorentz transformation and the resulting equations are shown to be the same.

On the other hand, Minkowski's emphasis is on the *geometric properties* of the theory, on those geometric entities which remain unchanged after the transformations, its *invariance*. Thus Minkowsky ensures the satisfaction of the principle of relativity by quite different means. The only structures allowed in constructing a theory are spacetime invariants.

Principle 4.18 (General Relativity Principle). *Inside Galileo's ship cruising with a generally curvilinear motion, no one can distinguish whether or not he is carried along in such a motion.*

The transformations between two frames of reference of GR are all arbitrary coordinate transformations differentiable with their inverse.

Principle 4.19 (General Covariance). *Under the transformations between two frames of reference of GR the equations of physics remain the same.*

For instance, inside a ship sailing along a circle, by applying a suitable coordinate transformation which introduces the inertial terms, namely the centrifugal and Coriolis' ones, the same law of motion stands, which is $F = ma$ for non-relativistic speed.

Note that: **spacetime reference changes play an analogous role with respect to coordinate changes.**

Weak Equivalence Principle (WEP$_a$): Freely falling particles move on timelike geodesics of spacetime.

Weak Equivalence Principle (WEP$_b$):   Freely falling particles take the path of maximal aging.

Weak Equivalence Principle (WEP$_c$):   All particles fall at the same rate in a gravitational field, independent of their mass, shape, dimension, composition, density, colour or anything else.

Strong Einstein Equivalence Principle (SEP$_\mathcal{E}$):   Any local physical experiment not involving gravity will have the same result if performed in a freely falling inertial frame[4] as if it were performed in the flat spacetime of special relativity[5].

Note that SEP$_\mathcal{E}$ is true provided that there is *no viscosity*. As a matter of fact, in presence of viscosity, the speed of free-falling depends on the shape of the falling body, if the external shape of the two falling bodies is the same then the speed depends on external dimensions if the external dimensions are the same (and so are the external shapes) than the speed of free-falling depends on the densities of the two bodies, namely from their masses.

The Happiest Idea of my Life (EHI):   As quoted by Einstein in 1915, *physics of curved spacetime must reduce over small regions to the physics of flat spacetime of special relativity.* Mathematically, this means that for each point p in the Universe (which is the *spacetime*) there exists a pair (N, φ) where N is an open neighbourhood of p and φ is a chart, which is a diffeomorphism between N and the Lorentz-Minkowsky spacetime ruled by the special relativity. In other words, the spacetime is *locally flat*, which means that using the notations just introduced, N is endowed with coordinates

$$\varphi \colon N \ni q \mapsto \bigl(t(q), x(q), y(q), z(q)\bigr) \in \varphi(N) \subseteq \mathbb{R}^4$$

so that the metric in these coordinates takes the canonical form $g_{\mu\nu} = \eta_{\mu\nu}$. Note that there is no guarantee that there is such a N which covers the whole manifold, and indeed the existence of *flat coordinates* carried by

---

[4] So that for each instant every point of the local frame is subject to the same gravity field (this is the requirement for localness), on the other hand, such a field of gravity has not to be *stationary*, in other words, it may vary from an instant to another.

[5] So, being in a flat spacetime *with special relativity* no gravity exists!

Definitions, Postulates and Principles

freely falling observers in which the physics is indistinguishable from that of special relativity is only *local*.

$SEP_\mathcal{E}$ and EHI are substantially equivalent so much that EHI could be assumed as another formulation of the Equivalence Principle $SEP_\mathcal{E}$. The equivalence principle stated as in EHI makes **General Relativity an extension of Special Relativity** to a curved spacetime, according to the principle of correspondence. This is probably the reason for Einstein's happiness.

In § 11.6. it will be shown in practice that by considering a too wide neighbourhood of Turin, namely a neighbourhood including the Plateau Rosà, whose space-distance from Turin consists in 90 km, the spacetime is no longer flat. Hence the need to take sufficiently small neighbourhoods.

Strong Equivalence Principle (SEP): *Any local physical experiment not involving gravity will have the same result if performed in an uniform gravitational field*[6] *or in a relativistic accelerated frame, such that each point is subjected to the same* 4−*acceleration*. Note that if the gravitational field is also *static* then the accelerated frame must be *uniformly* accelerated. Besides, note that under this last hypothesis, the magnitude of the gravitational field is intended to be the same as the magnitude of the four-acceleration. Anyway, four-acceleration and gravitational field's directions are anti-parallel.

Last but not the least, note that the above mentioned *relativistic acceleration* which, for each fixed instant has to be equal in each point of the frame, can be caused by any force in nature: electromagnetic, strong or weak.

A very useful consequence of SEP will be deduced beyond in § 9.4.

Proposition 4.20. *$SEP_\mathcal{E}$ imply WEP_c.*

Proof. The procedure is carried out with a reductio ad absurdum. Suppose an experiment consists in dropping a hammer and a feather together, from the same initial height, without air resistance, e.g. towards the surface of the Moon. Suppose to analyze the experiment in the reference frame of the lunar surface so that in such a frame both hammer and feather don't reach

---

[6]The hypothesis of *uniformity* regarding the gravitational field has its counterpart in the hypothesis of *locality* in $SEP_\mathcal{E}$.

the regolith at the same time, causing two different gravity acting on them, say $g_h \neq g_f$. Now, image to perform the same experiment in a free-falling inertial frame. So, assuming SEP$_\mathcal{E}$, it will have the same result as if were performed in the flat spacetime, without any gravity! In other words, in such a special frame hammer and feather do not differentiate each other since neither g_h nor g_f exists at all. Hammer and feather will be aligned, parallel to the lunar surface at each instant until they touch lunar regolith. Assuming the so-called *principle of excluded middle*, this is an evident contradiction. □

Principle 4.21 (Cosmological Principle). *Viewed on a sufficiently large scale Universe is homogeneous and isotropic. For more details see § 9.12.*

Principle 4.22 (Principle of Maximal Aging – PMA). *A particle that is not influenced by external forces follows the path that gives the largest possible proper time. For more details see § 11.4.*

Chapter 5

Where It All Began: The Light Clock

The light clock experiment designed by Einstein is presented, accompanied by intuitive and simple images. The invariance of the distances perpendicular to the motion will be analyzed together with the procedure necessary to measure the time intervals in the experiment, in order to then be able to deduce the famous equations of time dilation and contraction of distances. The angle that the light ray forms with the perpendicular to the direction of motion in the reference of the train tracks will also be studied.

Keywords: Light Clock, Time Dilation, Space Contraction, Aberration

One, perhaps the first one of Einstein's many dreams will be presented below. It is not an achievable experiment on Earth, due to the absence of any gravity field ... after all it is a dream, isn't it? So, imagine switching off the gravity and thinking about a train, moving along a straight rail at uniform speed, denoted by v_β in Figures 5.1 and 5.3.

Before analyzing Einstein's light clock and so the most famous problem concerned with times it is needed an important lemma regarding distances perpendicular to the direction of motion.

5.1. Perpendicular Distances Do Not Change

Perpendicular distances to the direction of motion do not undergo any modification, neither dilation nor contraction. In order to give complete proof of this assertion, it is possible to pursue a *reductio ad absurdum*.

Imagine a train stationary inside a straight gallery whose height coincides exactly with respect to train's one. Now, suppose the train is moving at uniform velocity inside the same gallery. Two eventualities are admissible: perpendicular dimension of a body in relative motion

(i) shrink

or, conversely,

(ii) enlarge.

In the first instance suppose to be inside the train, then by applying postulate no. 1 of Special Relativity train's height does not change. On the other hand, assuming hypothesis *(i)* the gallery's height contracts so that train does not pass through the gallery anymore. From the point of view of the gallery, which, differently from the train is fixed with respect to itself, the gallery's height does not change while, again by *(i)* train's height shrinks, which permits the train to pass through the gallery even more. This is a manifest contradiction since by the principle of the excluded middle a train can pass or not throw a gallery, independently from where the point of view of the observatory is set.

Analogously, the second instance supposes to be inside the train. Postulate 1 ensures that the train's height does not change, on the other hand, by assuming hypothesis *(ii)* the gallery's height enlarge so that train passes even more so through the gallery. Again, suppose to be steady in the gallery analyzing the train moving inside the gallery itself: obviously, the gallery's height does not change, on the other hand by assuming *(ii)* trains height enlarges so that train does not pass throw the gallery anymore, which is again an evident contradiction.

Therefore both the hypothesis *(i)* and *(ii)* must be discarded so that the initial assertion is logically proved.

Where It All Began: The Light Clock 49

5.2. The Physical Apparatus

On the train's floor, there is a laser gun which is positioned perpendicular to the floor itself. On the train's top, exactly over the gun, a plane mirror is positioned and intercepts every incoming beam of light reflecting it on the floor, where a photoelectric cell is positioned, exactly next to the gun. Contextually to the reception of the returning ray of light, a counter is increased and the gunshot another ray, and so on.

5.3. Two Systems of References

In the following we will be concerned with two systems of references:

(S_t) the first one coincident with the floor of the train, see Figure 5.1.

(S_r) the second one coincident with the rail, see Figure 5.3.

Who Is Really Moving with Respect To?

The two references S_t and S_r are *perfectly symmetric*. As a matter of fact S_t, namely the train, can say without fail that train is stationary and rails are moving on the left. On the other hand, correspondingly, S_r - the rails - can obviously say to be fixed with the train moving on the right.

However, anyone could anyway argue that it is the train that is obviously moving along the rails and *not* vice versa that rails are moving on the left along train wheels! This problem will be taken up in § 5.7..

5.4. Train's Time

Let's denote with $\Delta\tau$ the interval of time recorded by watch β in S_t between the two events consisting *in shooting the ray of light performed by the gun* and *receiving the same ray of light*, executed by the photoelectric cell, see Figure 5.1. In other words, $\Delta\tau$ is the quantum of time beaten by the train's watch β.

Is the interval of time between the two events consisting in shooting the ray (starting event) and receiving it (ending one) *proper*? It depends: surely yes in the frame of reference coincident with the train's floor since the two above mentioned events are determined in the same position, namely the train's floor, besides all the instants between them are measured in the same position, namely the train's floor. On the other hand, obviously, it is not a proper interval of time in the reference coincident with the ray of light, indeed, although the two events are determined in the same position, namely the train's floor, it is not true that all the instants between them are measured in the same position.

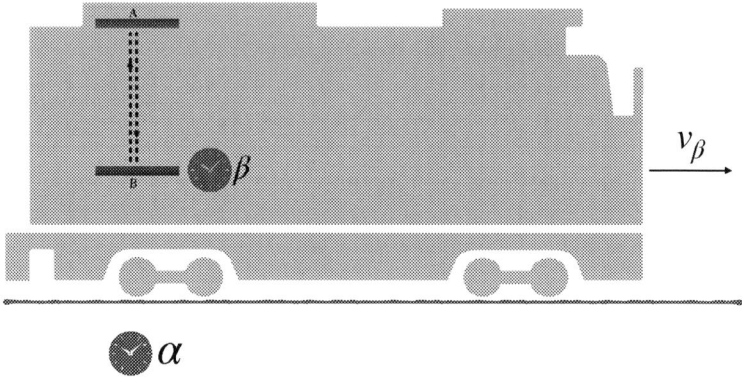

Figure 5.1.

5.5. Rail's Time

What about the corresponding time measured by the watch α belonging to the reference S_r? If it is denoted by Δt, standing the definition of proper time, with respect to the system of reference S_r it is not, evidently, a proper interval of time. As a matter of fact, the interval Δt separates two distinct instants which occur in *two different points in S_r*.

How Is It Possible to Measure It?

Referring to Figure 5.2 suppose that, while moving at speed v, at the starting moment, the laser gun, in addition, to shot a ray of light towards the train's ceiling fires simultaneously a ray of light directed towards the rail track, impressing it in a point P. Similarly, suppose the returning ray of light impresses also the rail track in Q. So it will result in turn that the space between the two points P and Q is indeed $v \cdot \Delta t$, from which Δt is indirectly determined by S_r by quoting the above mentioned space by v.

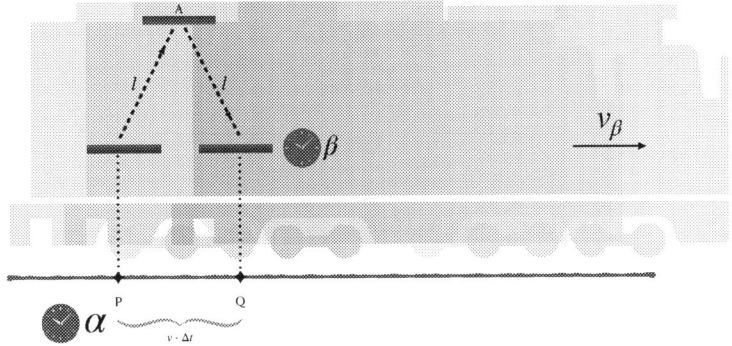

Figure 5.2.

Straightness of Trajectory

Note that the ray's of light trajectory is a straight line, which is a line with no curvature (so, in particular, no parabolic-hyperbolic motion in here!). As a matter of fact, it is *straight* since straight is the solution's trajectory of the equation of motion $\vec{r}'' \equiv \vec{0}$.

5.6. The Role Played by the Pythagorean Theorem

Standing the Special Relativity Principle, invoking the Postulate no. 1 of Special Relativity, all the physics inside the train is indistinguishable from the physics in the rail's reference frame. So:

$$\overline{AH} = c \cdot \frac{\Delta \tau}{2}. \tag{5.1}$$

Invoking the lemma concerned with the invariance of perpendicular distances, see § 5.1., the distance $\overline{AH}$ computed in S_t as in (5.1) is indeed the same as measured in S_r, so the following reasoning hold and in particular (5.4) stands.

Referring to Figure 5.3, ABB' is an isosceles triangle since $\ell := \overline{AB} = \overline{AB'}$. Besides, by construction, $A\widehat{H}B$ is a right angle. The following equations holds:

$$\ell = c \cdot \frac{\Delta t}{2} \tag{5.2}$$

$$\overline{BB'} = v \cdot \Delta t \quad \leadsto \quad \overline{BH} = v \cdot \frac{\Delta t}{2}. \tag{5.3}$$

Applying Pythagorean theorem to the triangle ABH it yields:

$$\overline{AB}^2 = \overline{AH}^2 + \overline{BH}^2 \tag{5.4}$$

which is equivalent to, by substituting:

$$\left(c \cdot \frac{\Delta t}{2}\right)^2 = \left(v \cdot \frac{\Delta t}{2}\right)^2 + \left(c \cdot \frac{\Delta \tau}{2}\right)^2 \tag{5.5}$$

which is

$$c^2 \cdot \Delta t^2 = v^2 \cdot \Delta t^2 + c^2 \cdot \Delta \tau^2$$

if and only if

$$\Delta t^2 = \frac{1}{1 - \left(\frac{v}{c}\right)^2} \cdot \Delta \tau^2$$

which, in turns, implies the following famous *time dilation equation*:

$$\Delta t = \frac{1}{\sqrt{1 - \left(\frac{v}{c}\right)^2}} \cdot \Delta \tau \tag{5.6}$$

where

$$\gamma = \gamma(v) := \frac{1}{\sqrt{1 - \left(\frac{v}{c}\right)^2}}$$

is the well known Lorentz gamma factor, which is, in particular, strictly greater than 1 for every $v \neq 0$. So, time is not equal for references in relative uniform motion, as a matter of fact, *rail's non-proper time is strictly greater than train's one*. At last, note that if light-train's speed v equals c than from equation 5.5 it follows that $\Delta \tau = 0$, namely the train *will not expertise time*.

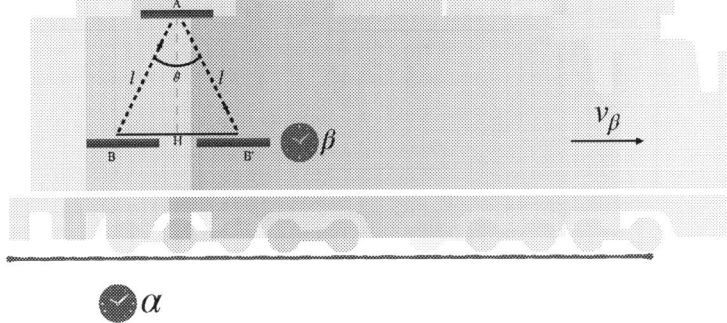

Figure 5.3.

5.7. Time Dilation?

Referring to § 5.3. the two frames S_t and S_r are indeed perfectly symmetric. As a matter of fact, by exchanging the experimental apparatuses belonging to train and rail, or better, by assembling another identical light clock on the rail, the symmetry claimed at the end of § 5.3. has been easily restored: no one will now object about the fact that watching α goes faster than β and vice-versa β goes faster than α (a watch α goes faster with respect to another one, say β if (and only if) the time beaten by it (α) is dilated with respect to the analogous one beaten by β). It seems to be a contradiction! Fortunately, it's only a paradox, as a matter of fact the two watches can be actually matched only if there is no relative motion between them, namely if and only if each one is stationary with respect to the other. While there is a relative motion between the two watches, both α and β will be able to say without fail that β and α respectively will go faster. So one of the two, α or β, will have to start chasing the other, irreparably damaging the symmetry.

Surely, during the acceleration phase of one of the two watches, see Figure 5.4, physics will undergo to laws different from the analogous one regarding the uniform relativistic motion which is already being determined. So, the accelerated watch will undergo correspondingly some perturbation that must certainly be taken into account. Suppose to have quantified these perturbations due to accelerations and so coherently take into account them into the final results (when the two watches are paired and no relative motion between them occurs). It will be finally possible to establish which of the two watches goes faster than the other? Absolutely not, since they are beating at the same time now. At least, will it be possible to establish which of the two watches had gone faster than the other, before the accelerating phase? Even this is not possible to know since the history of the watches is completely unknown and so it is not known which of the two watches was marking a lower hour than the other ... In other words it does not make any sense to ask which of the two watches goes faster than the other, simply because they are following two different *line of Universe* (see further § 8.1.).

Figure 5.4.

5.8. Distance Contractions

Starting from the equation $\Delta t = \gamma \Delta \tau$ it will be derived an analogous relation for distances. In order to pursue this target let's denote by $\Delta \sigma$ the wagon's length measured inside the wagon by a ruler which is fixed with respect to the wagon itself. So, by definition, $\Delta \sigma$ is a proper distance with respect to S_t. On the other hand, the analogous *indirect measure* performed by S_r (see § 5.9.) namely the wagon's length performed staying fixed in S_r, which is in relative motion with respect to the wagon, will be denoted by Δs. Here $\Delta \tau$ denotes the time measured staying fixed in S_r where a light clock has been previously assembled. Correspondingly Δt denotes the time measured by S_t. Standing these premises, since $\Delta t = \gamma \Delta \tau$ it results:

$$\frac{\Delta s}{\Delta \tau} = v = \frac{\Delta \sigma}{\Delta t}$$
$$\frac{\Delta s}{\Delta \tau} = \frac{\gamma \Delta s}{\Delta t}$$

which imply $\frac{\Delta \sigma}{\Delta t} = \frac{\gamma \Delta s}{\Delta t}$ and so the famous equation for *distance contractions*:

$$\Delta \sigma = \gamma \Delta s \iff \Delta s = \frac{\Delta \sigma}{\gamma}. \tag{5.7}$$

5.9. Indirect Relativistic Measures

Suppose someone is in a train moving of rectilinear uniform motion and he wants to measure the width of a building he is seeing from the window, parallel to the train rail. His equipment consists only of a stopwatch and a laser gun. So, pointing the laser perpendicular to the window he lets the stopwatch start as soon as he sees that the laser beam intercepts the building and correspondingly stops the timer as soon as the laser intercepts the end of the building. If $\Delta \tau$ denotes the time elapsed, the indirect relativistic measure of the building will be $\Delta s = v \cdot \Delta \tau$, where v denotes the relative speed between the train and the landscape.

5.10. Criticism to the Contraction of Distances

The following paragraph is aimed to put in evidence, as per time dilation, that also the so famous *contraction of distances* is not completely correct terminology.

Imagine having two identical trains, say T_1 and T_2, each one with its own light clock, moving uniformly in opposite rectilinear direction with relative speed v. Remember, no gravity and possibly trains to move in free space. Standing these hypotheses each train will be able to affirm that the other is moving with respect to oneself, so that, by the equation 5.7, each train will conclude that the other is shrunk. In other words, as per time dilation, the situation is perfectly symmetric. The only way to break this symmetry would be to establish with certainty which train is really moving with respect to the other. But, is it really possible standing the fact that trains are moving in interstellar space without gravity and any need of rails where trains' wheels roll? Under these hypotheses, it is absolutely impossible to break the perfect symmetry, so that each train will be able to affirm that the other is moving with respect to oneself, hence each train will be able to conclude that the other is shrunk.

5.11. Light's Climb Rate

In the rail's frame, being the speed of the oblique ray of light equal to c, since its horizontal component is v, the perpendicular component $c_\perp$ from H to A must satisfy the Pythagorean theorem $c^2 = c_\perp^2 + v^2$, if and only if

$$c_\perp = \sqrt{c^2 - v^2} = c\sqrt{1 - \left(\frac{v}{c}\right)^2} = \frac{c}{\gamma}.$$

So, the perpendicular component of the speed of light seen by S_r is different from c. As a matter of fact, this result can be obtained in another way by observing that:

$$c_\perp = \frac{\Delta s_{rail}}{\Delta t_{rail}} = \frac{\Delta \sigma_{train}}{\Delta t_{rail}} = \frac{\Delta \sigma_{train}}{\gamma \Delta \tau_{train}} = \frac{\Delta \sigma_{train}}{\gamma \Delta t_{train}} = \frac{c}{\gamma}$$

where

- Δs_{rail} denotes the the vertical path of the light seen by S_r

- $\Delta \sigma_{train}$ denotes the same path seen by S_t, hence it is a proper distance. Since Δs_{rail} is perpendicular to the direction of motion, by § 5.1. it results that $\Delta s_{rail} = \Delta \sigma_{train}$

- Δt_{rail} denotes the time measured in S_r necessary for the ray of light to reach B starting its motion in A (see Figure 5.3)

- $\Delta \tau_{rail}$ denotes the corresponding time measured in S_r

- the changing in notation $\Delta \tau_{train} = \Delta t_{train}$ is intended to put in evidence that, in the frame of the ray of light, what is denoted by $\Delta \tau_{train}$ is indeed a non-proper interval of time (see also the last paragraph of § 5.4.).

5.12. Aberration Angle

The angle $\widehat{BAH} = \frac{\theta}{2}$ formed by the velocities $\vec{c}$ and $\vec{c}_\perp$ is said to be the *aberration angle*, its value is given by:

$$\tan\left(\frac{\theta}{2}\right) = \frac{v}{c_\perp} = \gamma\frac{v}{c}.$$

Another effect of this phenomenon consists in the aberration of photons coming from stars. As a matter of fact, generally, stars lie in a direction perpendicular to the ecliptic, that is the plane of Earth's revolution around the Sun. Because of Earth motion, or equivalently, stars motion, they appear to lie in a slightly different direction than it would appear if stars were in a frame at rest relative to Sun. Anyway, although the aberration effect is very small, $\tan(\theta/2) \approx 10^{-4}$, in two opposite positions belonging to the ecliptic (which are two points separated by six months) the same star at the same time of the terrestrial day appears unequivocally in two slightly different positions. This was enough for Bradley (1725) to conclude the finiteness of light speed. Indeed, if the propagation of light were instantaneous ($c = \infty$) according to a non-relativistic treatment of the subject there would be no aberration:

$$\tan\left(\frac{\theta}{2}\right) = \frac{v}{c} \to 0$$

so that it would be $\theta = 0$.

For a sake of clarity, taking advantage of the Figure 5.3, B and B' can be thought of as the two above mentioned points in which the star would appear in two opposite positions belonging to the ecliptic.

A very common analogy consists in considering the apparent direction of the rain dripping on a train window (here the moving train is the rotating Earth and the rain plays the analogous role of a ray of light).

Chapter 6

The Lorentz Transformations

> *This chapter presents Lorentz transformations, a fundamental tool for understanding the structure of space and time in Special Relativity. After their demonstration, their properties are analyzed, in order to then deriving from them the formulas of the dilation of times and the contraction of distances. Starting from them the formula of the relativistic composition of velocities is then demonstrated, and the fundamental relation of the Invariant Interval is presented.*

Keywords: Lorentz transformations, Composition of velocities, Invariant Interval

Sometimes, thinking about Lorentz transformation our mind flies to something concerned with *coordinate transformations*. Then the doubt comes in: perhaps Lorentz transformations regards *transformations between two references*, maybe in uniform rectilinear motion one with respect to the other.

As a matter of fact, both the points of view are shareable, true, since, as stated in § 4.5., spacetime *reference changes* play an analogous role with respect to *coordinate changes*. Effectively, given a *system of coordinates*, since it consists of a rigid axis whose mutual orientation does not vary with time, see for example Figure 6.1, it determines in a natural way a *system of* reference. Vice versa, given a *system of reference* it determines naturally a *system of coordinates* since, by definition, a system of reference is a rigid body. In the previous chapter, space contraction has been deduced from

Figure 6.1.

time dilation. Now it is possible to show that space contraction implies all Lorentz equations.

6.1. Lorentz Space-Equation and Its Inverse

Referring to Figure 6.2 it results

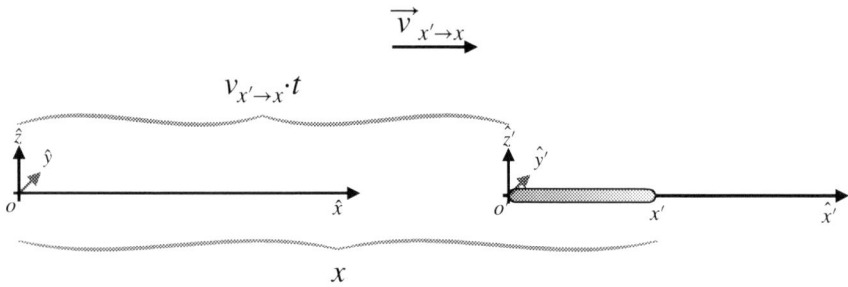

Figure 6.2.

$$\boxed{\frac{x' - 0'}{\gamma} = x - v_{x' \to x} \cdot t} \quad (6.1)$$

so

$$x' = \gamma(x - v_{x' \to x} t). \quad (6.2)$$

Note that $v_{x' \to x}$ is here intended to be *positive* since $\vec{v}_{x' \to x}$ has the same direction of the frame of coordinate underneath by x. Conversely, by exchanging the role of x with x', that is, considering x which is moving in

The Lorentz Transformations

opposite direction with respect to x' with velocity $\vec{v}_{x \to x'}$ with respect to x':

$$\frac{x-0}{\gamma} = -v_{x \to x'} \cdot t' + x' \tag{6.3}$$

in which note that $v_{x \to x'} \cdot t'$ is negative (since it is $v_{x \to x'}$) and also negative is x', so that $-v_{x \to x'} \cdot t'$ is positive and $0 < -v_{x \to x'} \cdot t' + x' < -v_{x \to x'} \cdot t'$.

So

$$x = \gamma \left(x' - v_{x \to x'} t' \right)$$

or, equivalently, standing that $v_{x \to x'} = -v_{x' \to x}$

$$x = \gamma \left(x' + v_{x' \to x} t' \right). \tag{6.4}$$

6.2. Lorentz Time-Equation and Its Inverse

Referring to Figure 6.3 in order to deduce Lorentz time-equation and its in-

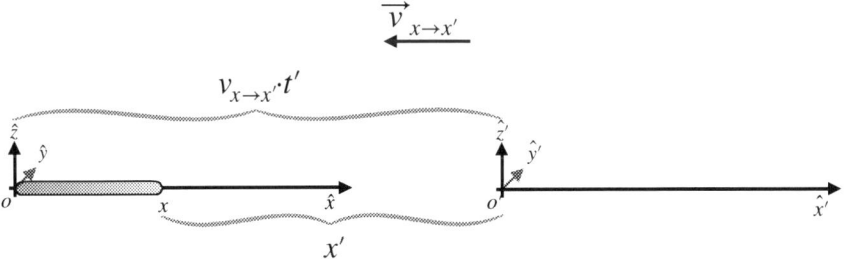

Figure 6.3.

verse, it's particularly convenient to make a system between equations 6.1 and 6.3. First of all note the following technicism:

$$\frac{1}{\gamma} - \gamma = -\left(\frac{v}{c}\right)^2 \gamma. \tag{6.5}$$

It results:

$$\begin{cases} \frac{x'}{\gamma} = x - v_{x' \to x} \cdot t \\ \frac{x}{\gamma} = x' - v_{x \to x'} \cdot t' \end{cases} \iff \begin{cases} x' = \gamma(x - v_{x' \to x} \cdot t) \\ x = \gamma(x' - v_{x \to x'} \cdot t') \end{cases}$$

from which it follows equivalently:

$$\begin{cases} \frac{x'}{\gamma} = \gamma(x' - v_{x \to x'} \cdot t') - v_{x' \to x} \cdot t \\ \frac{x}{\gamma} = \gamma(x - v_{x' \to x} \cdot t) - v_{x \to x'} \cdot t' \end{cases}$$

From the first one it can be easily obtained t as a function of t', conversely in the second one t' versus t, so that respectively

$$t = \gamma\left(t' - \frac{v_{x \to x'}}{c^2} x'\right) \tag{6.6}$$

or equivalently

$$t = \gamma\left(t' + \frac{v_{x' \to x}}{c^2} x'\right) \tag{6.7}$$

and

$$t' = \gamma\left(t - \frac{v_{x' \to x}}{c^2} x\right). \tag{6.8}$$

In order to remember both the direct and inverse equations for space and time, it is particularly convenient to take into account that all the equations are of the form:

$$\cdots = \gamma(\cdots - v \cdots)$$

where v denotes the speed of a certain frame S with respect to another T, such that all the physical quantities between parenthesis in the second member are referred to T.

Finite Difference Lorentz Equations

Starting from (6.4) and (6.7), the following finite difference relations (which can be immediately generalized to differential ones substituting Δ with d operator) will be very useful in the sequel. In particular, standing the constancy of the relative velocity it follows that:

$$\Delta x = \gamma \Delta\left(x' + v_{x' \to x} t'\right) = \gamma\left(\Delta x' + v_{x' \to x} \Delta t'\right) \tag{6.9}$$

and similarly

$$\Delta t = \gamma\left(\Delta t' + \frac{v_{x' \to x}}{c^2} \Delta x'\right). \tag{6.10}$$

Note that $\Delta x'$ denotes $x'(t'_2) - x'(t'_1)$ and analogously the other differences.

6.3. Properties of Lorentz Equations

The following preamble prepares the ground to show that time dilation and space contraction are equivalent to Lorentz equations for space and time. First of all note that the equations for space contraction

$$\Delta\sigma = \gamma\Delta s$$

and time dilation

$$\Delta t = \gamma\Delta\tau$$

describe physical models with *two* frame of reference. On the other hand the equations 6.9 and 6.10 describe the motions of physical models with *three* frame of reference, which are:

- (S) the first frame of reference describes its own quantities measured through the coordinate system x, t

- (S') the second frame of reference describes its own quantities measured through the coordinate system x', t'

- (S_P) the third frame of reference does not perform measurements, but its motion is described by the frame of reference S and S'. In Figure 6.4 it is represented by the point P.

These *three* systems of reference can be either a *physical body or imaginary elements*, what is necessary is that they are *inertial* reference systems, which move with uniform rectilinear motion with respect to each other.

Deriving Time Dilation and Space Contractions

Standing the above premises, in order to deduce time dilation and space contraction equations from Lorentz transformations it is necessary to eliminate one of the three reference systems. So, translate S' until the two points O' and P coincide. In such a manner the two reference systems S' and S_P will coincide. It follows that $\Delta x'$, which quantifies the difference in coordinates between P and O' equals 0.

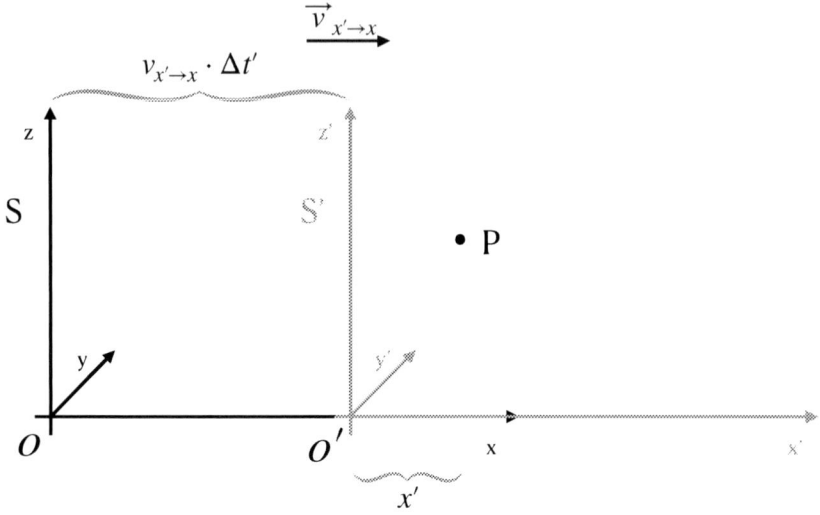

Figure 6.4.

Length Contraction

Substituting $\Delta x' = 0$ in equation 6.9 it yields

$$\Delta x = \gamma \left(\Delta x' + v_{x' \to x} \Delta t' \right) = \gamma \, v_{x' \to x} \Delta t'$$

where $v_{x' \to x} t' = v_{x' \to x} \Delta t'$ is the spatial distance measured in the reference system $S' \equiv S_p$, which moves with respect to S, and which therefore measures a contracted distance. It follows that:

$$\Delta x = \gamma v \Delta t' \iff v \Delta t' = \frac{\Delta x}{\gamma}$$

is equivalent to the equation 5.7, where $\Delta \sigma$ plays the role of Δx and Δs plays the role of $v \Delta t'$.

Time Dilation

Substituting $\Delta x' = 0$ in equation 6.10 it yields

$$\Delta t = \gamma \left(\Delta t' + \frac{v \Delta x'}{c^2} \right) = \gamma \Delta t'$$

which is equivalent to equation 5.6 in which $\Delta\tau$ plays the role of $\Delta t'$.

Deriving the Composition of Velocities

I do not know what reasoning led Einstein to postulate the constancy of the speed of light. Anyway, if the propagation of light were instantaneous then for every v it would result $c + v = \infty + v = \infty = c$. On the other hand, since the speed of light is finite (there are lots of experiments proving this, starting from J. Bradley experiment about annual-revolution aberration, see § 5.12.) Einstein postulated an analogous relation which further experiments proved to be written in the intimate nature of the Universe:

$$c \oplus v = c$$

where "$\oplus$" will be something but not the ordinary "+" operation.

Starting from equations 6.2 and 6.8, namely

$$x' = \gamma(x - v_{x' \to x} t).$$

and

$$t' = \gamma\left(t - \frac{v_{x' \to x}}{c^2} x\right),$$

and dividing them it results:

$$v_{p \to x'} = \frac{x'}{t'} = \frac{\gamma(x - v_{x' \to x} t)}{\gamma\left(t - \frac{v_{x' \to x}}{c^2} x\right)} = \frac{\frac{x}{t} - v_{x' \to x}}{1 - \frac{v_{x' \to x}}{c^2} \frac{x}{t}} = \frac{v_{p \to x} - v_{x' \to x}}{1 - \frac{v_{x' \to x} \cdot v_{p \to x}}{c^2}}$$

that is

$$\boxed{v_{p \to x'} = \frac{v_{p \to x} - v_{x' \to x}}{1 - \frac{v_{p \to x} \cdot v_{x' \to x}}{c^2}}} \qquad (6.11)$$

which is equivalent to its inverse:

$$\boxed{v_{p \to x} = \frac{v_{p \to x'} + v_{x' \to x}}{1 + \frac{v_{p \to x'} \cdot v_{x' \to x}}{c^2}}} \qquad (6.12)$$

6.4. Invariant Interval

Given the spacetime interval $\left(\Delta t, \frac{\Delta x}{c}\right)$, where, for sake of uniformity, spaces have been homogenized with respect to time (dividing canonically by the speed of light), the so-called *Lorentz-Minkowsky's norm* is the unique distance in the spacetime (see Figure 6.5) which is *invariant* under Lorentz transformations, and so, as demonstrated above, which takes into account both time dilation and space contraction. It is defined by:

$$\left\|\left(\Delta t, \frac{\Delta x}{c}\right)\right\|_{LM} := \sqrt{(\Delta t)^2 - \left(\frac{\Delta x}{c}\right)^2}. \qquad (6.13)$$

Equation 6.14 will rigorously show that the distance above defined is in-

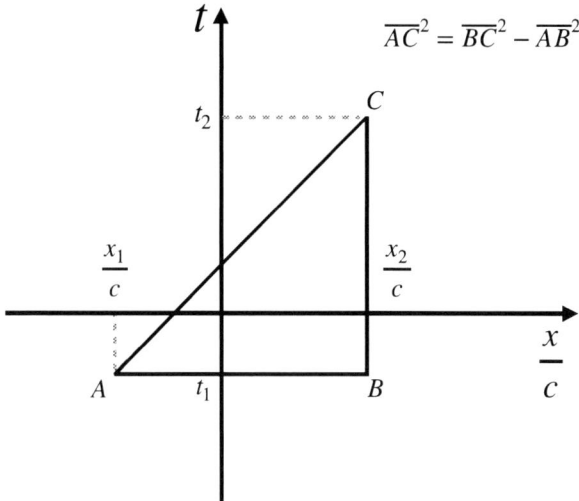

Figure 6.5.

variant under Lorentz transformations. As a matter of fact, it is possible to prove that any other norm in spacetime which is invariant under Lorentz transformation coincides with Lorentz-Minkowsky's norm, in other words, this property characterizes the Lorentz-Minkowsky's norm.

The Lorentz Transformations

Starting from the equations

$$\begin{cases} \Delta x' = \gamma(\Delta x - v_{x' \to x}\Delta t) \\ \Delta t' = \gamma\left(\Delta t - \frac{v_{x' \to x}}{c^2}\Delta x\right) \end{cases} \iff \begin{cases} \Delta x = \gamma(\Delta x' - v_{x \to x'}\Delta t') \\ \Delta t = \gamma\left(\Delta t' - \frac{v_{x \to x'}}{c^2}\Delta x'\right) \end{cases}$$

according to (6.13), since $\gamma^2\left(1 - \left(\frac{v}{c}\right)^2\right) = 1$, a direct computation shows that (here v stands for $v_{x \to x'}$)

$$(\Delta t)^2 - \left(\frac{\Delta x}{c}\right)^2 = \gamma^2\left(1 - \left(\frac{v}{c}\right)^2\right) \cdot (\Delta t')^2 - \gamma^2\left(1 - \left(\frac{v}{c}\right)^2\right) \cdot \left(\frac{\Delta x'}{c}\right)^2 = (\Delta t')^2 - \left(\frac{\Delta x'}{c}\right)^2 \tag{6.14}$$

proving that the Lorentz-Minkowsky's norm of the spacetime interval is invariant under Lorentz transformation.

In conclusion, time dilation implies space contraction, which implies Lorentz transformation, which implies, on the one hand, both time dilation and space contraction, on the other hand, composition of velocities and the so famous above discussed invariant interval.

The following diagram (6.6) attempts to summarize the sections of the chapter and how they relate to each other:

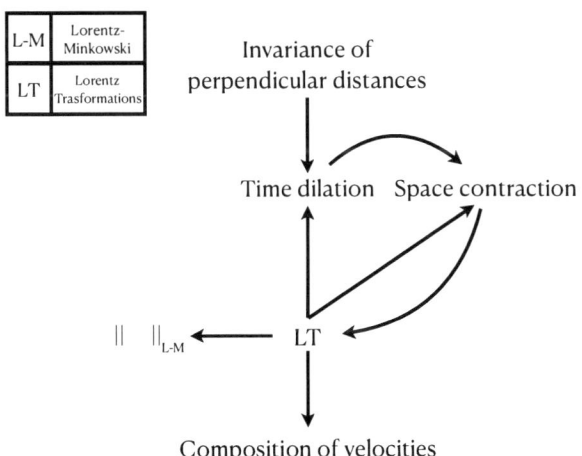

Figure 6.6.

Chapter 7
Simultaneity and Causality

Galilean relativity, in the 17th century, explained for the first time how velocities and spaces are not absolute; but in 1905 Albert Einstein proved that time was not absolute as well, pushing even further the meaning of measurements in physics and our understanding of the world. In this chapter the goal is to deeply explain the nature of events and their properties in relation to space and time, analyzing the consequences of Einstein's work. The concept of simultaneity is therefore defined and analyzed in the relativistic field. Subsequently, the analyzes of the famous paradoxes concerning this topic are presented. Finally, the concept of causality and chronological order is studied, with interesting implications.

Keywords: Simultaneity, Causality, Chronological Order, Train Paradox, Car and Garage Paradox

7.1. Introduction

First of all, it is important to keep this definition in mind:

Definition. Two events $\mathcal{E}_1$ and $\mathcal{E}_2$ which occur respectively on the points P_1 and P_2 are said to be *simultaneous* if the light beams emitted by the two points in the instants of the occurrences of the two events themselves arrive at the *midpoint M* at the *same time*.

In fact, referring to the second postulate of special relativity, it is known that the speed of light, in empty space, is the same for all observers. Therefore:

$$\begin{cases} c = \dfrac{\overline{P_1M}}{\Delta t_1} \\ c = \dfrac{\overline{P_2M}}{\Delta t_2} \\ \overline{P_1M} = \overline{P_2M} \end{cases} \Rightarrow \begin{cases} \dfrac{\overline{P_1M}}{\Delta t_1} = \dfrac{\overline{P_2M}}{\Delta t_2} \\ \overline{P_1M} = \overline{P_2M} = d_M \end{cases} \Rightarrow \dfrac{d_M}{\Delta t_1} = \dfrac{d_M}{\Delta t_2} \Rightarrow \Delta t_1 = \Delta t_2$$

(7.1)

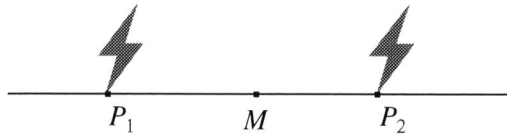

Figure 7.1.

7.2. Einstein's Train Paradox

Einstein's train paradox is the most famous example of how the simultaneity of events in general relativity becomes relative. In this paradox two light bolts hit two different points P_1 and P_2 in the train track at the same instant, as shown in Figure 7.2.

There are two observers, one watching the train from the ground at a given point M and another one sitting on the train at a given point M', assuming that, in the track's reference, they are initially equidistant from the points P_1 and P_2.

Observer M is hit by the light of the light bolts at the same time, but observer M first perceives the light source it is going against by moving.

This paradox though may give the idea that the events are relative only by an optical illusion, but again, the events observed by M' not only appear optically different, for that observer they actually happened at different times.

Simultaneity and Causality

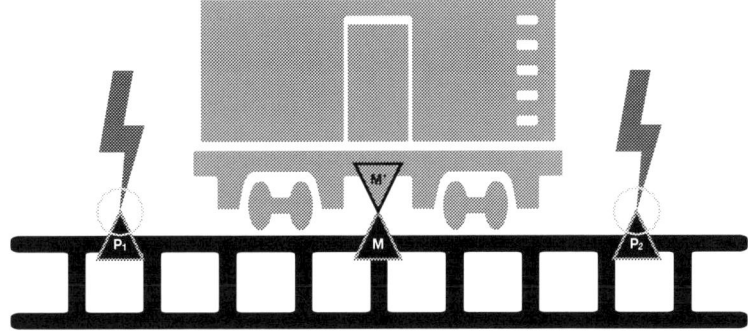

Figure 7.2.

Train Track's Reference Frame

As already said, in the train track reference the two light bolts fall at the same time in the instant, which will be called t_0, when the points M and M' are overlapped and equidistant from the points P_1 and P_2, so that

$$\overline{P_1M} = \overline{P_2M} = \overline{P_1M'} = \overline{P_2M'},$$

as shown in Figure 7.2.

The light emitted by the two light bolts starts travelling heading towards M and M'. While M stays still in his reference, M' instead travels ahead the light emitted by the P_1's light bolt and away from the light emitted by the P_2's light bolt.

This results in the observer M' seeing before the P_2's light bolt and after the P_1's light bolt, as shown in Figures 7.3 and 7.5; the observer M instead, as expected from equation 7.1, sees the two light bolts at the same time (Figure 7.4).

Train's Reference Frame

Now the same situation in the train's reference frame is proposed.

In this reference, the train, and so the observer M', stays still and the rails, and so the observer M, travel at a relative speed v to the left, as shown in Figure 7.6.

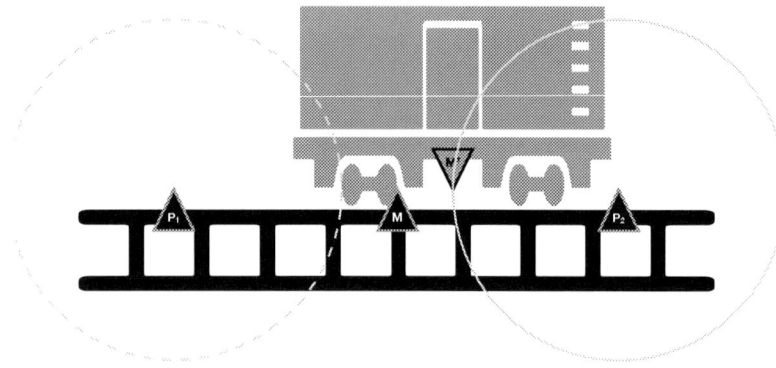

Figure 7.3.

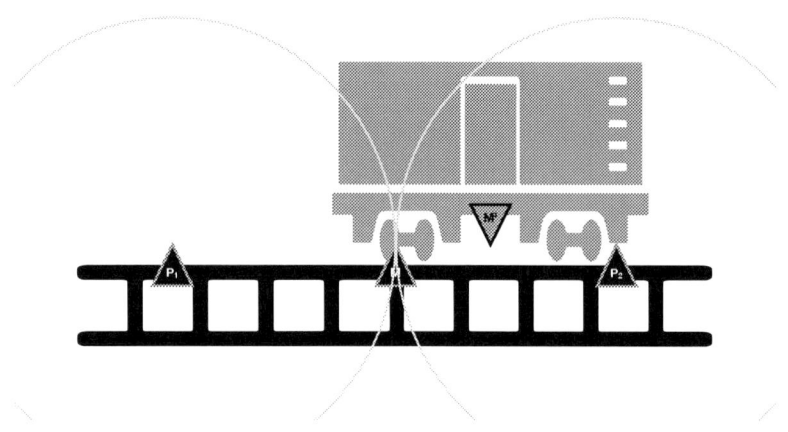

Figure 7.4.

At a certain instant, which will be called $t'_{0,2}$, the P_2's light bolt happens in the train's reference frame and its light starts spreading towards M and M' (Figure 7.7).

Due to the *Second Postulate of Special Relativity* the points P_2 and P_1 will be the centre of their respective rays of light's propagation only in the instant they emit them.

Now the P_2's light bolt's light will spread until it reaches M' (Fig-

Simultaneity and Causality

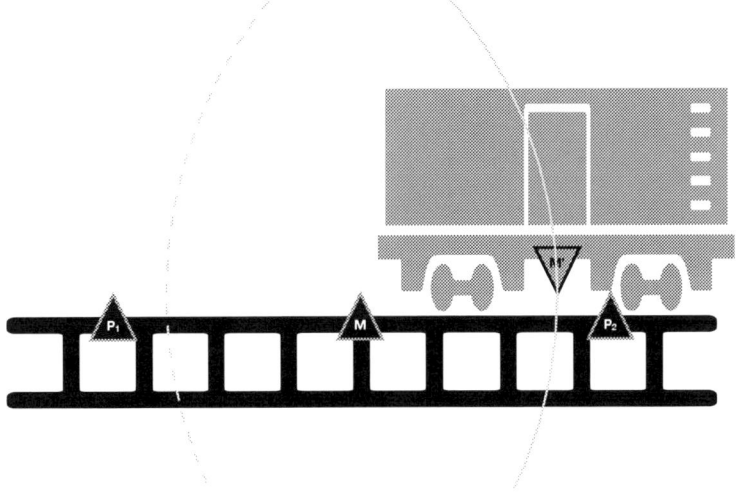

Figure 7.5.

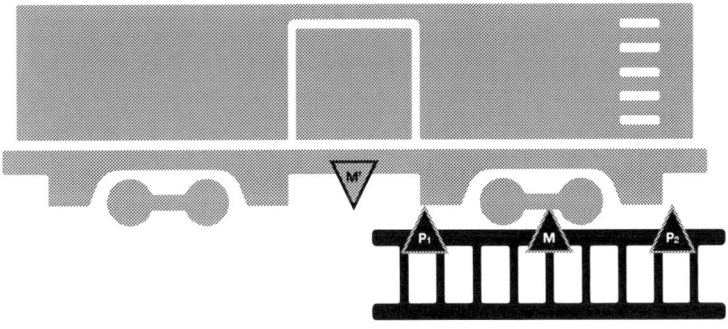

Figure 7.6.

ure 7.8).

At a certain instant, which will be called $t'_{0,1}$, the P'_1's light bolt happens in the train's reference frame (Figure 7.9).

Then the P_2's light bolt's and the P'_1s light bolt's lights will reach M together.

Finally the P'_1s light bolt's light will reach M'.

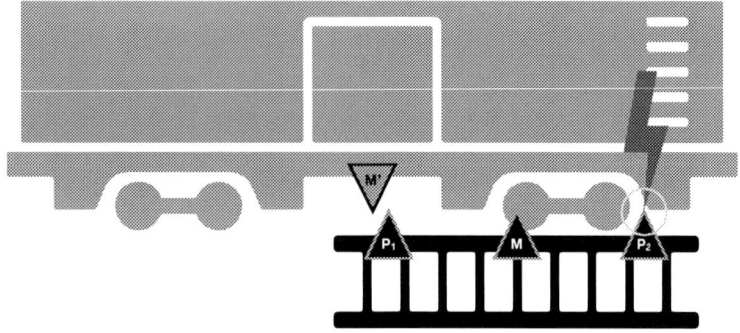

Figure 7.7.

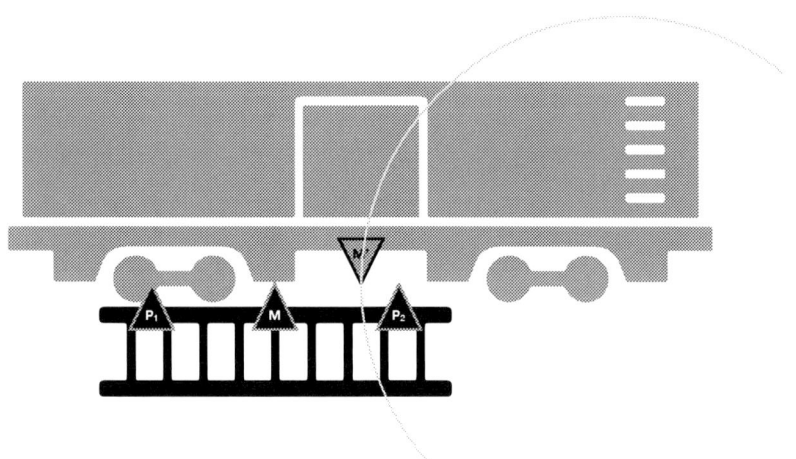

Figure 7.8.

So it is possible to conclude that also in the train's reference frame the two events won't be seen simultaneously by the observer M', while they will be seen simultaneously by the observer M.

If this statement couldn't be true it would mean that non only the timing of the events, in different reference frames, is different but also that the existence of the single event itself would be different, meaning that an

Simultaneity and Causality

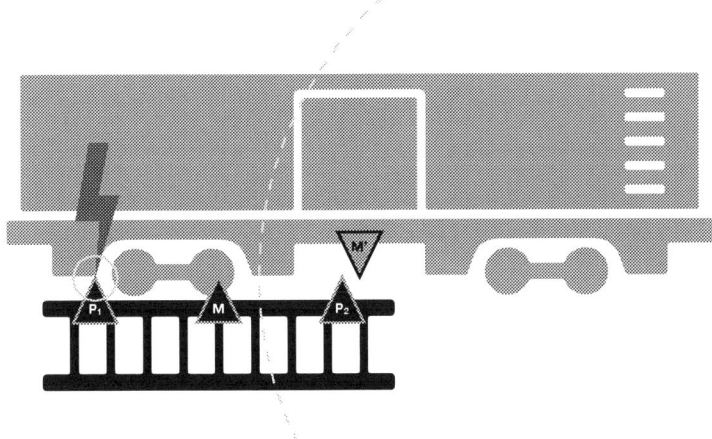

Figure 7.9.

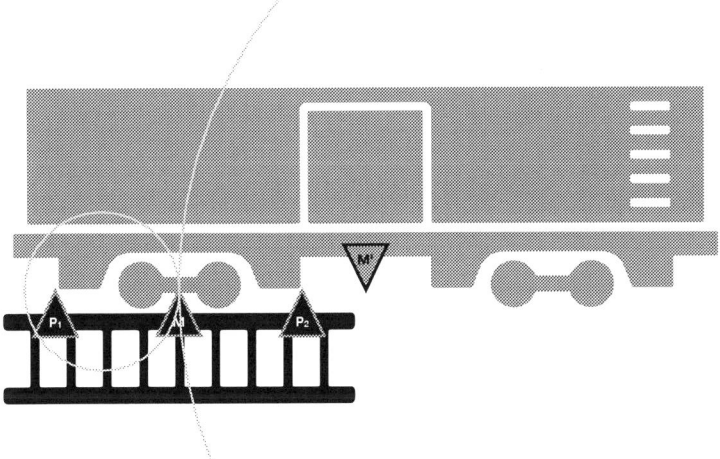

Figure 7.10.

event which, for example, exists for the train doesn't exist for the rails, and that is impossible.

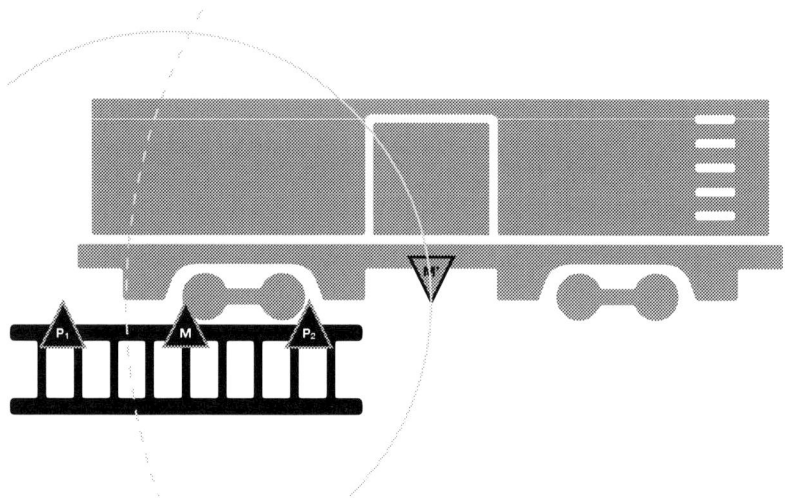

Figure 7.11.

The order of these events may variate in relation with the relative speed and the length of the train but the event shown in Figure 7.10 will always happen between the events shown in Figure 7.8 and Figure 7.11.

Time Interval Measured by the Train Track

Let's now calculate the time interval Δt that exists between the two events in the **train's track reference frame**.

In the train's track reference a system of coordinates positive towards the right with his origin in M is fixed. In the first place it will be necessary to determine the equation of motion, in the reference of the train track, of the ray of light coming from P_2:

$$X_2(t) = -ct + d$$

The equation of motion, in the reference of the train track, of the ray of light coming from P_1 is:

$$X_2(t) = ct - d$$

where d is the distance $\overline{P_2M} = \overline{P_1M}$ which is the distance measured in the train track reference frame.

Simultaneity and Causality

Finally, the equation of motion of the point M' in the reference of the train track is:
$$X_{M'}(t) = vt.$$
Therefore, the ray of light coming from P_2 reaches M' if:
$$-ct_2 + d = vt_2 \iff t_2 = \frac{d}{c+v}$$
and the ray of light coming from P_1 reaches M' if:
$$ct_1 - d = vt_1 \iff t_1 = \frac{d}{c-v}.$$

Now let's calculate the interval $\Delta t = t_1 - t_2$ that passes between the instant t_2, when the P_2's light bolt's light reaches M' in the train track's reference frame, and the instant t_1, when the P_1's light bolt's light reaches M' in the train track's reference frame.

$$\Delta t = \frac{d}{c-v} - \frac{d}{c+v} = \frac{d(c+v) - d(c-v)}{c^2 - v^2} = \frac{dc + dv - dc + dv}{c^2 - v^2}$$

$$= \frac{2dv}{c^2(1 - \frac{v^2}{c^2})} = \gamma^2 \frac{2dv}{c^2}$$

So in the train track's reference frame the two events happen in the train with a Δt:
$$\Delta t = \gamma^2 \frac{2d \cdot v}{c^2}.$$

Time Interval Measured by the Train

Let's consider now the same situation in the **train's reference frame**.

Using the Lorentz Equations for time 6.8 it is possible to determine the previously named $t'_{0,2}$ and $t'_{0,1}$ which are respectively the instants when the P_2's lighbolt and the P_1's lighbolt happen in the train's reference frame:

$$t'_{0,2} = \gamma \left(t_{0,2} - \frac{x_2 v}{c^2} \right)$$

$$t'_{0,1} = \gamma \left(t_{0,1} - \frac{x_1 v}{c^2} \right)$$

It is known that both light bolts happen at the same time in the train track's reference frame, so $t_{0,2} = t_{0,1} = 0$.

It is also known that the x_2 and x_1 coordinates of both the light bolts are respectively equal to $+d$ and $-d$, in the train track's reference frame.

So it is possible to write:

$$t'_{0,2} = \gamma\left(0 - \frac{dv}{c^2}\right) \iff t'_{0,2} = -\gamma\frac{dv}{c^2}$$

$$t'_{0,1} = \gamma\left(0 - \frac{-dv}{c^2}\right) \iff t'_{0,1} = \gamma\frac{dv}{c^2}$$

Now it will be computed the interval $\Delta t' = t'_{0,1} - t'_{0,2}$ between the two events in the train's reference frame.

$$\Delta t' = \gamma\frac{dv}{c^2} - \left(-\gamma\frac{dv}{c^2}\right) = \gamma\frac{2dv}{c^2}$$

It is now possible to start noticing some similarities between Δt and $\Delta t'$. First of all it is necessary to remember that $\Delta t'$ is the interval between the two events in the train's reference frame and it is equal to, due to the equation 7.1, the interval between the instants in which P_2's light bolt's light and P_1's light bolt's light reach the observer M', also measured in the train's reference frame; Δt instead is the same interval but measured in the train track's reference frame. So that $\Delta t'$ is a proper time while Δt is a non-proper time.

Given that the equation 5.6 of time dilation can be exploited:

$$\Delta t = \gamma \Delta t' \iff \gamma^2 \frac{2dv}{c^2} = \gamma\left(\gamma\frac{2dv}{c^2}\right) \iff 1 = 1.$$

This means that, actually, two events that are simultaneously for a reference frame are not simultaneous for another reference frame which is moving relative to the first one. Besides the interval between the light emitted by the two events reaching the moving observer measured in the first reference frame is actually the dilated time interval, measured in the moving reference frame, with which the two events happen.

7.3. Train Paradox - Light Sensors' Variation

At this point an equivalent version of the previous paradox is provided, useful for understanding the phenomenon. Instead of thunderbolts, two light sensors will be used, located on the train tracks at points A and B. When the front wheel of the train passes over point B and the rear wheel passes over point A, the light sensors emit a beam of light which is directed towards point M, equidistant from both ends. Let's see what happens in the two different reference frames:

- the train track reference frame
- the train reference frame.

They are presented in the following pages.

Train Track's Reference Frame

Now the analysis is brought into the train track's reference frame. The train moves to the right with a relative speed v. Its wheels, which will be called A' and B' respectively, are approaching points A and B, as shown in Figure 7.12.

It is important to remember that, in the track reference,

$$\overline{AM} = \overline{BM} = \overline{A'M'} = \overline{B'M'},$$

where M' is the midpoint of $\overline{A'B'}$. The train passes over points A and B and, as stated above, the wheels of the train, or points A' and B', will coincide for an instant with points A and B of the tracks. In this instant the light sensors emit their signal directed towards the point M of the rails, as shown in Figures 7.13, 7.14, 7.15 and 7.16.

As it is possible to see, the two events, which are the emission of the two light signals, are simultaneous in the reference system of the train tracks. In fact, the space that the light will have to travel will be the same in both cases and therefore the time taken to travel this space will also be the same. Ultimately the two events satisfy the definition of simultaneity presented in § 7.1.

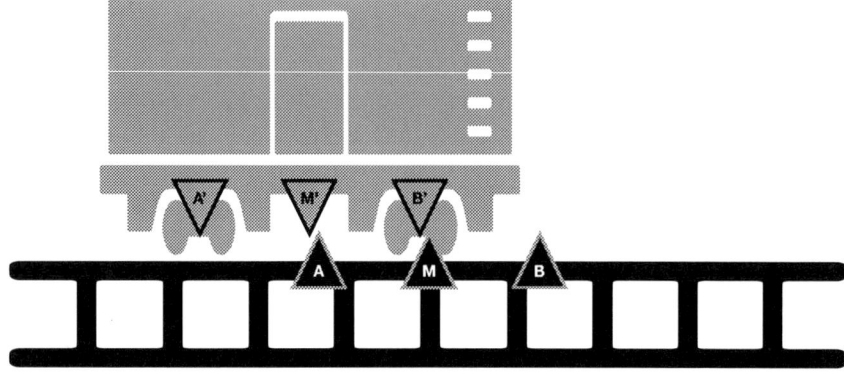

Figure 7.12.

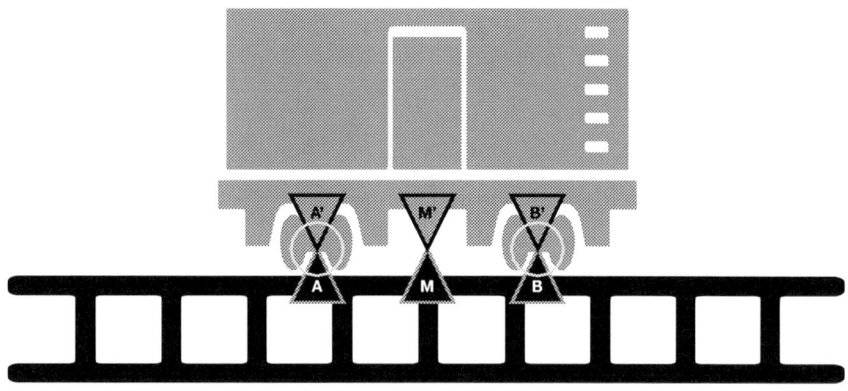

Figure 7.13.

Train's Reference Frame

Now it is necessary to imagine the same situation in another reference frame, the train's reference frame.

In this reference the train is stationary, and it is the rails that move at relative speed v to the left, as shown in Figure 7.17.

Simultaneity and Causality

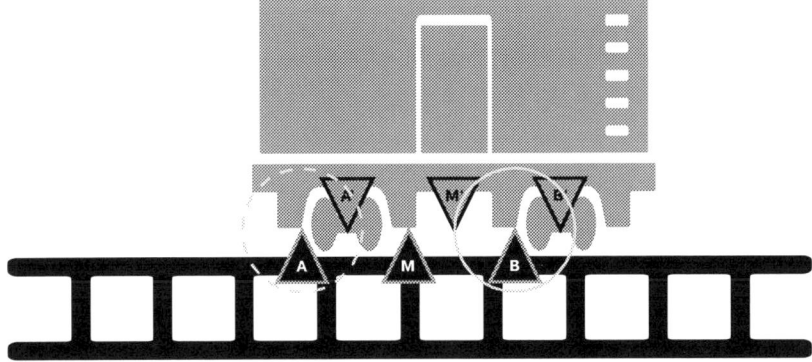

Figure 7.14.

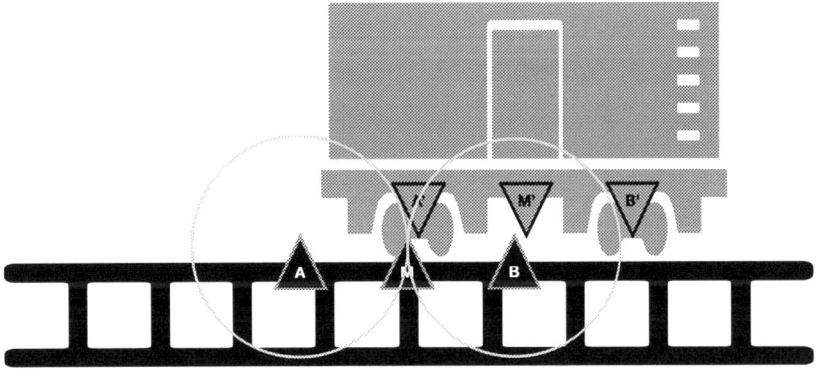

Figure 7.15.

Furthermore, the rails themselves will be shortened, due to the equation of space contraction. Therefore the equality $\overline{AM} = \overline{BM} = \overline{A'M'} = \overline{B'M'}$ will no longer be true. There will not be an instant where points A and B will coincide with points A' and B' respectively.

Instead, initially, it will be the front wheel of the train B' to pass over the point B, triggering the light sensor which will send the signal to the

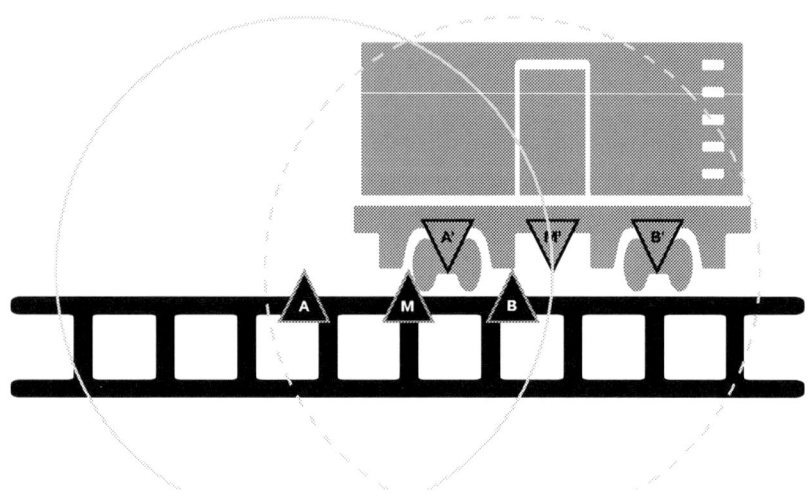

Figure 7.16.

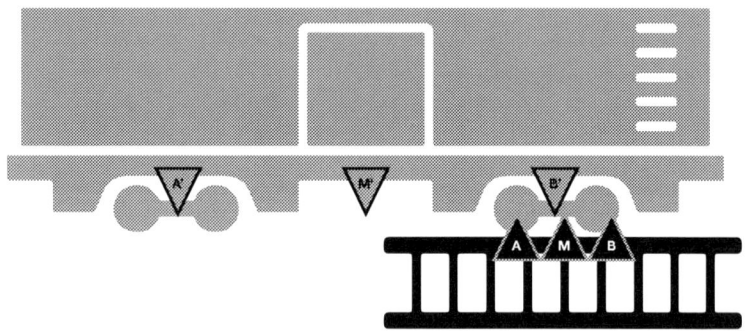

Figure 7.17.

centre M', as shown in Figure 7.18.

The signal will be propagated until it reaches the centre, as shown by Figure 7.19.

After a certain time interval Δt also the rear wheel A' will pass over the point A, triggering the light sensor which will send the signal to the center

Simultaneity and Causality

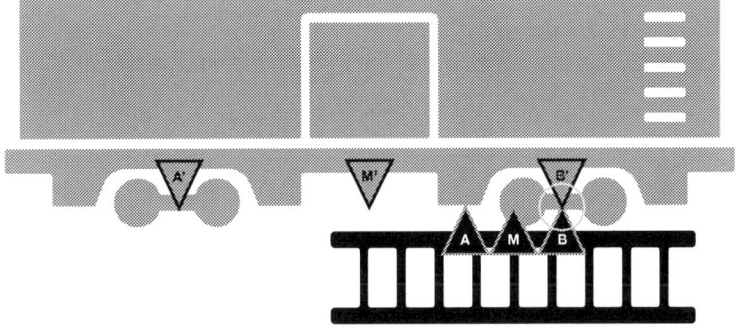

Figure 7.18.

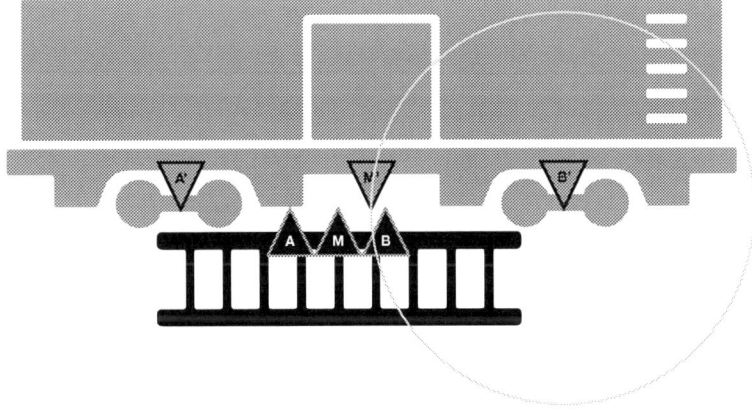

Figure 7.19.

M', as shown in Figure 7.20.

The signal will be propagated until it reaches the center, Figures 7.21, and 7.22.

So it is possible to conclude that in the train's reference frame the two events do not respect the definition 7.1 of simultaneity (while they will be seen simultaneously by M) as the two signals transmitted by the light sensors (and so by the points A' and B', see the figures) arrive at different

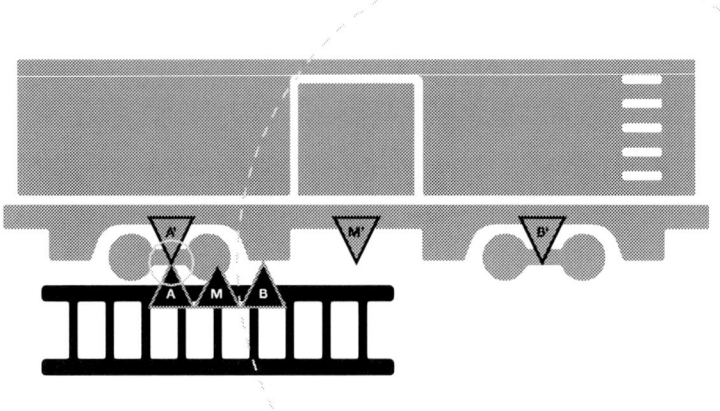

Figure 7.20.

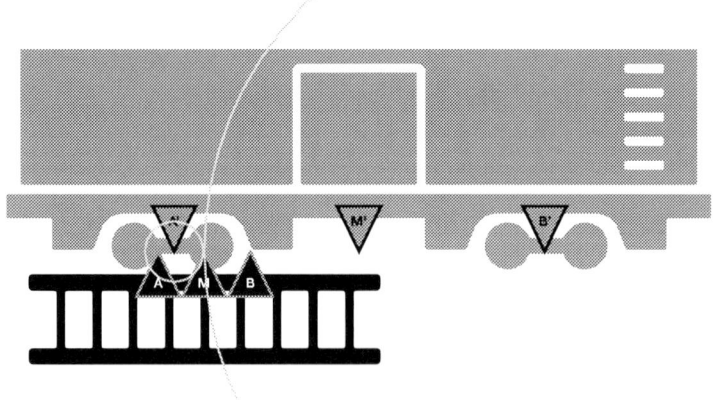

Figure 7.21.

instants in the midpoint M'.

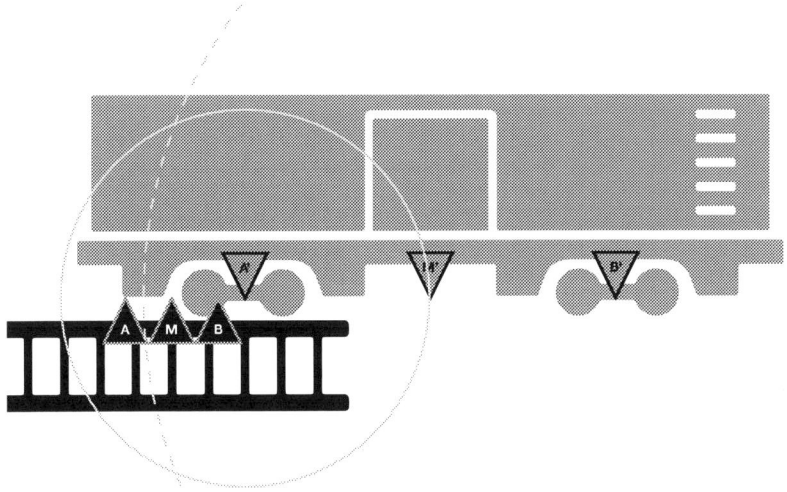

Figure 7.22.

Time Interval Measured by the Train Track

Let's now calculate the time interval Δt that exists between the two events in the **train track's reference frame**.

First of all, in the reference of the train track, it will be necessary to determine the equation of motion of the ray of light coming from A'. Fixed in the train track reference a system of coordinates positive towards the right with its origin in $A \equiv A'$ it is obtained:

$$X_{A'}(t) = ct.$$

Similarly, the equation of motion of the ray of light coming from B' is:

$$X_{B'}(t) = -ct + L$$

where L is the length of the train in the train track's reference frame, and

$$\mathcal{L} \stackrel{\text{def}}{=} \gamma L \qquad (7.2)$$

is the proper length of the train (i.e. the length measured in the train's reference frame).

Finally, the equation of motion in the same reference frame of the point M' is:
$$X_{M'}(t) = vt + \frac{L}{2}.$$
Therefore, the ray of light coming from A' reaches M' if and only if
$$ct = vt + \frac{L}{2} \iff t \stackrel{\text{def}}{=} t_1 = \frac{L}{2(c-v)}.$$
And the ray of light coming from B' reaches M' iff
$$-ct + L = vt + \frac{L}{2} \iff t \stackrel{\text{def}}{=} t_2 = \frac{L}{2(c+v)}.$$
The first time obtained has been called t_1 and the second time t_2. If the two events were simultaneous it is possible to prove that $t_1 = t_2$ and therefore $t_1 - t_2 = 0$. On the other hand, following the previous arguments, time interval $\Delta t \neq 0$ is expected to be found. Let's calculate it.
$$t_1 - t_2 = \frac{L}{2(c-v)} - \frac{L}{2(c+v)} = \frac{Lc + Lv - Lc + Lv}{2(c^2 - v^2)}$$
$$= \frac{2Lv}{2(c^2 - v^2)} = \frac{Lv}{c^2\left(1 - (\frac{v}{c})^2\right)} = \gamma^2 L \frac{v}{c^2}.$$
It has been shown that in the train track's reference frame the two events are not simultaneous and that the time interval between the first event and the second one is equal to
$$\Delta t = \gamma^2 L \frac{v}{c^2} = \gamma L \frac{v}{c^2}.$$
Also, studying physics in the train reference frame the two events have not to be simultaneous, indeed it will show that they are separated by the obviously contracted time interval consisting in $\Delta t' = \frac{\Delta t}{\gamma}$.

Time Interval Measured by the Train

Now it is possible to consider the same situation in the **train's reference frame**. In this reference the train is stationary and the train track moves to the left. It follows that (see figure 7.18) the track is contracted and also

Simultaneity and Causality

is $\overline{AB}$, which become $\overline{AB} = \frac{L}{\gamma} = \frac{L/\gamma}{\gamma} = \frac{L}{\gamma^2}$ where, as stated in equation 7.2, is put $\overline{A'B'} = L$. Let's consider as the starting instant the figure 7.18. In the train reference frame, the sensor in A' emits its light beam with a delay (with respect to the analogous one emitted by B') equal to the time necessary to A necessary to cover at speed v the distance from A (at its starting position, when B coincides with B') to A', i.e.

$$\frac{\overline{A'B'} - \overline{AB}}{v} = \frac{L - L/\gamma}{v} = \frac{L - L/\gamma^2}{v} = \frac{L}{v}(1 - 1/\gamma^2) = \frac{L}{v}\frac{v^2}{c^2} = \gamma L \frac{v}{c^2}.$$

Fixed a system of coordinates positive towards the right with its origin in A', the equation of motion of the points B', A' and M' in the train's reference frame can be written.

The equation of motion of the ray of light coming from B' is:

$$X_{B'}(t') = -ct' + L.$$

The equation of motion of the ray of light coming from A' is:

$$X_{A'}(t') = \begin{cases} 0 & \text{if } t' \leq \gamma L \frac{v}{c^2} \\ (t' - \gamma L \frac{v}{c^2})c & \text{else.} \end{cases}$$

Finally, the equation of motion in the same reference frame of the point M' is:

$$X_{M'}(t') \equiv \frac{L}{2}.$$

Therefore, the ray of light coming from A' reaches M' if

$$\left(t' - \gamma L \frac{v}{c^2}\right)c = \frac{L}{2} \iff t'c = L\frac{v}{c} + \frac{L}{2} \iff t' \stackrel{\text{def}}{=} t'_1 = \frac{L(2v+c)}{2c^2}.$$

On the other hand, the ray of light coming from B' reaches M' if

$$-ct' + L = \frac{L}{2} \iff t' \stackrel{\text{def}}{=} t'_2 = \frac{L}{2c}.$$

It follows that the interval between the two events, i.e. the interval between the instants in which the photon coming from B and the photon coming from A reach M', measured in the train's reference frame equals to

$$\Delta t' = t_1 - t_2 = \frac{L(2v+c)}{2c^2} - \frac{L}{2c} = L\frac{v}{c^2} = \gamma L \frac{v}{c^2}, \tag{7.3}$$

as expected.

By solving this, it is demonstrated, again, that actually two events that are simultaneously for a reference frame are not simultaneous for another reference frame which is moving relative to the first one. Besides, it has been shown that the interval between the light emitted by the two sensors reaching the moving observer measured in the track reference frame is actually the dilated time interval measured in the moving reference frame.

Using the Lorentz equation for time, $\Delta t' = \gamma(\Delta t - v\Delta x/c^2)$, it is possible to determine the previously relationship 7.3. In fact, it is known that the two events, that are the emission of the two light signals, are simultaneous in the reference system of the train tracks (as already noticed in 7.3.) and so $\Delta t = 0$.

It is also known that in the train track's reference frame the x_1 and x_2 coordinates of the light signal coming from A and B are equal to 0 and L respectively, and so $\Delta x = x_1 - x_2 = 0 - L = -L$. It follows, by using Lorentz equation of time, that $\Delta t' = \gamma(-v(-L)/c^2) = \gamma Lv/c^2$, as shown in (7.3).

7.4. Car and Garage Paradox

The *car and garage paradox*, also known as *ladder paradox*, is another common example of the relativity of events, this model is perfectly equivalent to the previous variation of the Einstein's train paradox, but in this case, only the triggering events of the two light sensors are considered. In this paradox a garage of proper length $\overline{AB}$, who sees a car moving at a speed such that the car's proper length $\overline{A'B'}$ is contracted to exactly $\overline{AB}$, since it moves at a relativistic speed towards the garage (Figure 7.23).

For the car though, the garage is shorter than the car, as in its reference frame $\overline{A'B'} > \overline{AB}/\gamma$ (Figure 7.24).

The Nature of the Paradox

The experiment consists of closing for an instant the front gate - equivalent to the left light sensor - and the back gate - equivalent to the right light sensor - of the garage at the same time, trying to close the car inside: from the garage reference frame by hypothesis $\overline{AB} = \overline{A'B'}$, so that the car perfectly fits inside the garage (Figure 7.25).

Simultaneity and Causality 89

Figure 7.23.

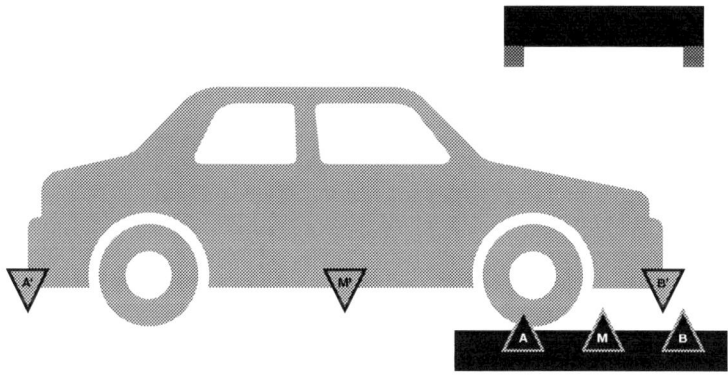

Figure 7.24.

But as described before, on the car reference frame the garage would be seen smaller than the car itself: in this case, it could be expected that the car will be hit by both gates, contradicting the events on the other frame leading to a reliability issue (Figure 7.26).

Figure 7.25.

Figure 7.26.

Solving the Paradox

First, it is necessary to understand that this situation is based on special relativity's contraction of distances, which is, in turn, the other side of the coin of the contraction of times. The problem in fact resides precisely in the transition from a reference frame to the other, and, since every reference frame has its own associated coordinate system, from a coordinate system

Simultaneity and Causality

to the other.

When switching between two coordinate systems the *Lorentz equations* are used, and, in this case, it is already known as a hypothesis how the contractions of the spaces between the two transform, but it is not yet possible to know what the temporal contractions are. In particular, it is not possible to simply assume from memory that the two-time coordinates of the events of the closing of the doors change in the same way; it is necessary to consider the possibility that they change in different ways, leading to a perceived non-simultaneity of the two, in the reference framework of the car.

The paradox is then solved by realising that for both observers, at a certain point in time, the head of the car will coincide with the front door, and the tail of the car will coincide with the back door.

The car reference perceives the events of the closing gates in different times: first, the front gate closes and opens instantly (Figure 7.27), then the car continues to travel until its tail surpasses the back gate, which closes only after it is left behind (Figure 7.28).

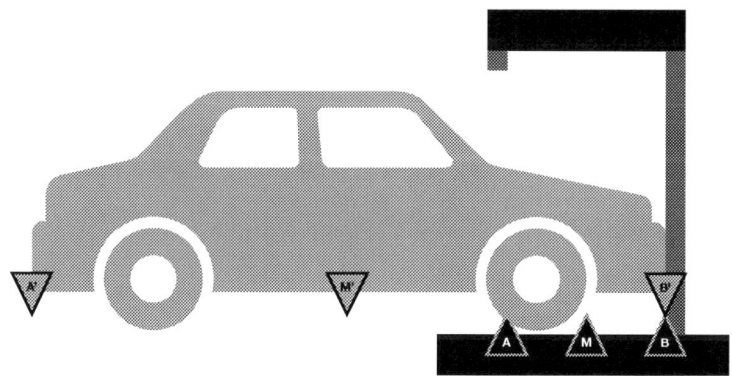

Figure 7.27.

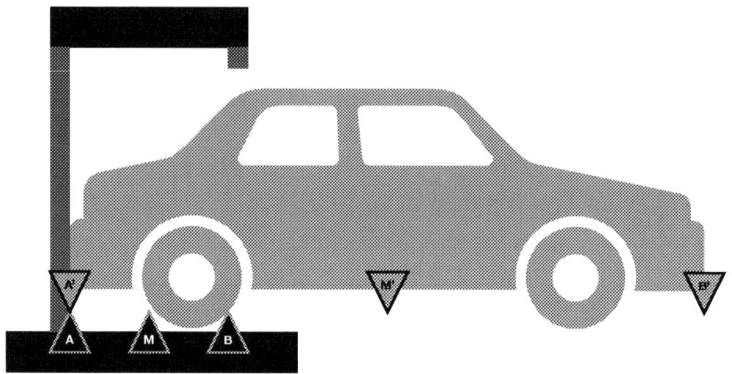

Figure 7.28.

Time Interval Measured by the Car

To determine the time variation $\Delta t'$ between the events shown in figures 7.27 and 7.28 the Lorenz transformations (equation 6.8) can be used to find the time of the two events t'_1 (Figure 7.27) and t'_2 (Figure 7.28) and subtract them. It is first assumed t_1 and t_2 (the corresponding time coordinates in the garage's frame) as 0 and consider x_1 and x_2 as the corresponding spacial coordinates of the back gate (points A and B in figure), with the origins of the coordinate system O set respectively on point M. It will also be assumed that the two coordinate systems are both oriented as the velocity vector of the car.

$$t'_1 = \gamma\left(t_1 - \frac{vx_1}{c^2}\right) = -\gamma\frac{vx_1}{c^2} = -\gamma\frac{v\left(-\frac{l}{2}\right)}{c^2} = \gamma\frac{vl}{2c^2}$$

$$t'_2 = \gamma\left(t_2 - \frac{vx_2}{c^2}\right) = -\gamma\frac{vx_2}{c^2} = -\gamma\frac{v\left(\frac{l}{2}\right)}{c^2} = -\gamma\frac{vl}{2c^2}$$

where l, as in 7.2., is the proper length $\overline{AB}$ of the garage.
In conclusion

$$\Delta t' = |t'_2 - t'_1| = \left|-\gamma\frac{vl}{2c^2} - \gamma\frac{vl}{2c^2}\right| = \gamma\frac{vl}{c^2}$$

Simultaneity and Causality 93

With this final equation, it is finally proved that the events are perceived by the two observers as stretched and staggered with direct proportionality to the relative velocity *v* and the distance between the event and the observer. In other words, for the moving observer, the more an event is distant from him, and the faster he goes, the more his time will be staggered from the other stationary observer.

7.5. Chronological Order

Under the same assumptions of section 7.3. will be now supposed that - when the front wheel of the train passes over point *B* and the rear one passes over *A* - the light sensor in *B* emits a beam of light with a delay quantified in Δt with respect to *A*, as shown in the following figures.

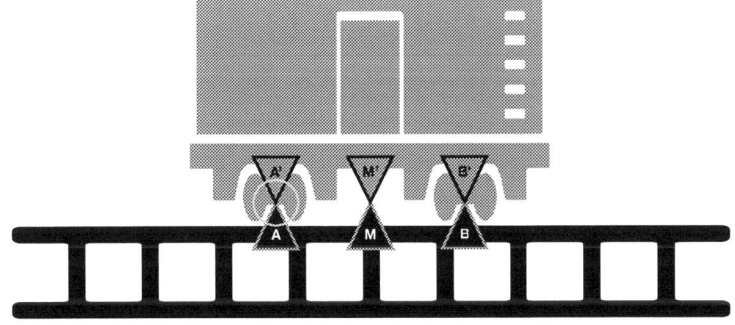

Figure 7.29.

Then, clearly, in the train's track reference frame, the signals will arrive in the midpoint *M* in the following order:

1. the signal coming from *A* (Figure 7.31)

and, after Δt,

2. the signal coming from *B* (Figure 7.32)

On the other hand, analyzing also in this case what happens in the train from the point of view of the track reference frame, it is possible to

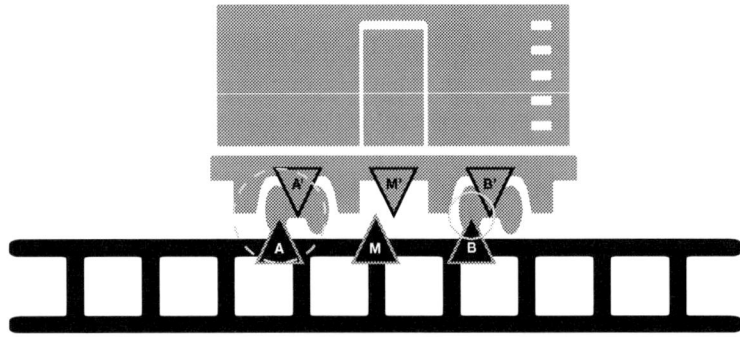

Figure 7.30.

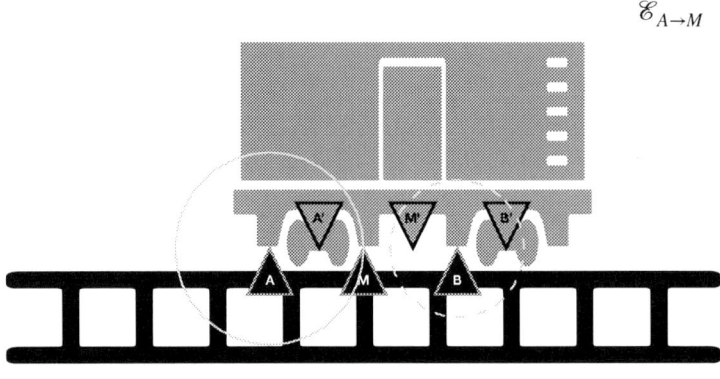

$\mathscr{E}_{A \to M}$

Figure 7.31.

conclude that the light beam coming from B' is anticipated of $\gamma^2 Lv/c^2$ with respect to the beam emitted by A', as stated in section 7.6. Thus, if Δt above is such that $\Delta t < \gamma^2 Lv/c^2$ then, in the train track reference frame, the signals emitted from A' and B' will arrive in the midpoint M' in the following order:

1. the signal coming from B (Figure 7.33)

and, after $\gamma^2 Lv/c^2 - \Delta t$,

Simultaneity and Causality

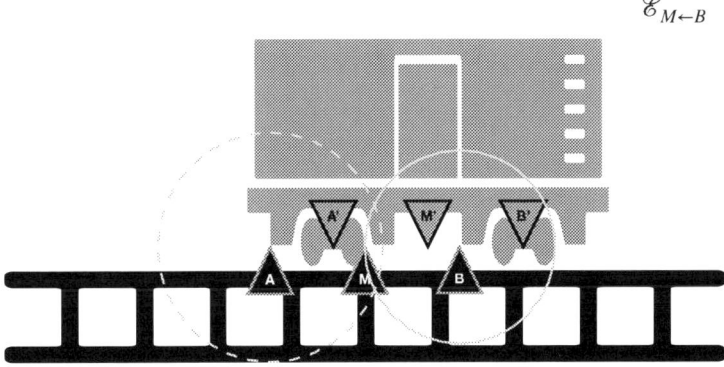

Figure 7.32.

2. the signal coming from A (Figure 7.34)

The chronological order in the two references is obviously reversed.

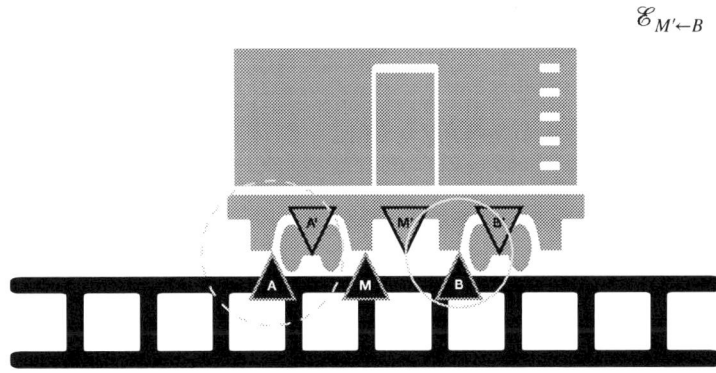

Figure 7.33.

Note that not only do the two light beams arrive at the midpoint M' reversed compared to the arrivals on M, in M' reference frame M' sees itself struck the two events in reversed order as well. This tells us that the events arranged on the temporal line of an observer (the events that occur in the

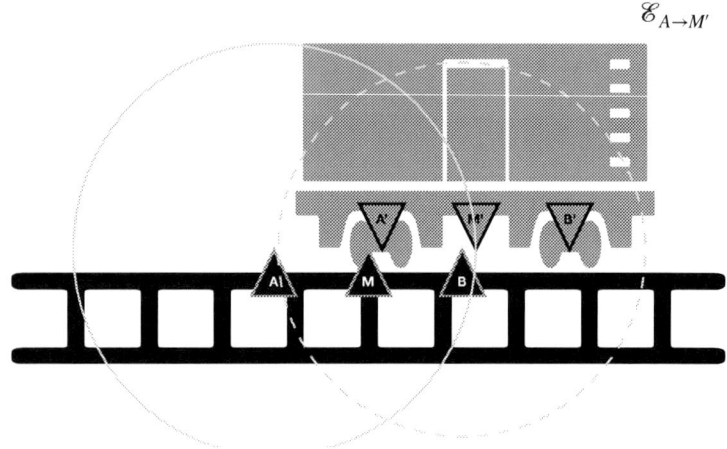

Figure 7.34.

points of space that coincide with that observer) maintain the same temporal order in any frame of reference; on the other hand, events that occur in generic points of spacetime can be arranged in an order that varies according to the chosen observer. From this, it is possible to conclude that, in general, both simultaneity and chronological order of events is dismissed.

7.6. Causality

If it is possible that times of events are actually reversible, what about the *principle of Causality* so important to philosophers? Can causes and effects really exchange with each other? In special relativity cause and effects cannot be inverted: the only reversible events are the ones separated through space, as seen in the previous paragraphs, since the mathematical component that allows such a thing is vx/c^2. When it comes to causality, however, pairs of events are considered to be related in two different ways:

- Proper events of a certain medium, such that an event happening to that object, will later have a consequence on the same object.

- Events happening on a certain medium, that will later have a consequence on another medium.

Simultaneity and Causality

The first case can be demonstrated as follows:

Proof. Consider two generic events $\mathcal{E}_1$ and $\mathcal{E}_2$ proper of a certain medium, and in that medium's timeline, they are ordered such that $t'_2 > t'_1$, where t'_2 is the proper time of $\mathcal{E}_2$ and t'_1 is the proper time of $\mathcal{E}_1$. Our goal it so prove that $t_2 > t_1$ as well, where t_2 and t_1 are the corresponding time coordinates of any other frame. To achieve such result will be used again the Lorentz equation of time, where x_1 and x_2 are the spatial coordinates of the two events in any chosen frame. Note that the position of $\mathcal{E}_1$ and $\mathcal{E}_2$ coincide with the position of the medium since the events are proper of it.

$$t'_2 > t'_1$$

$$\gamma\left(t_2 - \frac{vx_2}{c^2}\right) > \gamma\left(t_1 - \frac{vx_1}{c^2}\right)$$

since γ is always greater than 0, it is possible to simplify it

$$t_2 - \frac{vx_2}{c^2} > t_1 - \frac{vx_1}{c^2}$$

$$t_2 > t_1 - \frac{vx_1}{c^2} + \frac{vx_2}{c^2}$$

$$t_2 > t_1 + \frac{v}{c^2}(x_2 - x_1)$$

considering that $\mathcal{E}_1$ and $\mathcal{E}_2$ coincide with the position of the medium, it is possible to see that

$$x_2 = x_1 + v\Delta t = x_1 + v(t_2 - t_1)$$

substituting it inside the previous equation

$$t_2 > t_1 + \frac{v}{c^2}(x_1 + v(t_2 - t_1) - x_1)$$

$$t_2 > t_1 + \frac{v}{c^2}(v(t_2 - t_1))$$

$$t_2 > t_1 + \frac{v^2}{c^2}(t_2 - t_1)$$

$$t_2 > t_1 + t_2\frac{v^2}{c^2} - t_1\frac{v^2}{c^2}$$

$$t_2 - t_2\frac{v^2}{c^2} > t_1 - t_1\frac{v^2}{c^2}$$

$$t_2\left(1 - \frac{v^2}{c^2}\right) > t_1\left(1 - \frac{v^2}{c^2}\right)$$

note that the term $1 - \frac{v^2}{c^2}$ is always greater than 0 and

$$t_2 > t_1$$

□

therefore causality is safe just because in every circumstance $v \leq c$.

The second case's demonstration can be developed similarly to the first one:

Proof. Let $\mathcal{E}_1$ be a generic event of spacetime and $\mathcal{E}_2$ it's consequence on a certain medium. In that medium's timeline, they are ordered such that $t'_2 > t'_1$, where t'_2 is the proper time of $\mathcal{E}_2$ and t'_1 is the non-proper time of $\mathcal{E}_1$. Again, x_1 and x_2 are the spacial coordinates of the two events in any chosen frame.

$$t'_2 > t'_1$$

$$\gamma\left(t_2 - \frac{vx_2}{c^2}\right) > \gamma\left(t_1 - \frac{vx_1}{c^2}\right)$$

$$t_2 - \frac{vx_2}{c^2} > t_1 - \frac{vx_1}{c^2}$$

$$t_2 > t_1 - \frac{vx_1}{c^2} + \frac{vx_2}{c^2}$$

Simultaneity and Causality

$$t_2 > t_1 + \frac{v}{c^2}(x_2 - x_1)$$

in this case x_2 can be expressed as follows:

$$x_2 = x_1 + w\Delta t = x_1 + w(t_2 - t_1)$$

where w is the velocity with which $\mathcal{E}_1$ is interfering with the medium, it could be a signal or an emitted object, anything that would imply the consequence $\mathcal{E}_2$. Note that w can't be faster than c.

$$t_2 > t_1 + \frac{v}{c^2}(x_1 + w(t_2 - t_1) - x_1)$$

$$t_2 > t_1 + \frac{v}{c^2}(w(t_2 - t_1))$$

$$t_2 > t_1 + \frac{vw}{c^2}(t_2 - t_1)$$

$$t_2 > t_1 + t_2\frac{vw}{c^2} - t_1\frac{vw}{c^2}$$

$$t_2 + t_2\frac{vw}{c^2} > t_1 - t_1\frac{vw}{c^2}$$

$$t_2\left(1 - \frac{vw}{c^2}\right) > t_1\left(1 - \frac{vw}{c^2}\right)$$

again, $1 - \frac{vw}{c^2}$ is always greater than 0, even if w gets close to $\pm c$:

$$t_2 > t_1.$$

□

7.7. Conclusion

Considering events separated throw space, that are not a consequence of each other, such as in 7.2., it is seen that the time of the two events $t_{0,1} = 0$ and $t_{0,2} = 0$, in the train's reference frame become respectively

$$t'_{0,1} = \gamma \frac{dv}{c^2}$$

and

$$t'_{0,2} = -\gamma \frac{dv}{c^2}$$

The time of the two events shifted in a way such that they are equal, but with opposite signs. It could be possible to think that since $t'_{0,2}$ shifts time with a minus sign then the moving observer actually went back in time, but this is not the case: the only thing that all frames of reference agree on is the occurrence of events, the shift of times is just a result of different alignments of the different observers.

In conclusion, simultaneity is not absolute, any observer perceives events with an order and a separation-related to his speed and distance from its measurements.

The event's time variation is explained by equation 6.8, with particular attention to the member

$$\frac{vx}{c^2}$$

In Einstein's relativity, the simultaneity principle becomes relative to the velocity of the observers. In spacetime all objects move to carry a different time, they all travel at the same speed (through spacetime), but not in the same direction.

The orientation of space and time gives us different simultaneity of events, not just by an optical illusion, but in an actual difference in perspective and perception. Two observers that move at different speeds have different definitions of space and time, which causes their spacetime axes to be different, and perceive events differently, but the causality principle remains valid.

Subpart B
Special Relativity: Kinematics

Chapter 8

Lorentz-Minkowsky's Spacetime

The goal of this chapter is to rigorously study Lorentz-Minkowsky spacetime, starting from its non-trivial and often erroneous representation. The concepts of Four-Position, Four-Velocity and Four-Acceleration are therefore defined and the reasons that lead to the division of a proper quantity with a non-proper one or vice versa are studied. Finally, after a brief reflection on the lines of the Universe, a glimpse at dynamics in the relativistic field is presented, focusing on the concepts of momentum, force and energy.

Keywords: Spacetime, Four-Position, -Velocity, -Acceleration, -Momentum, -Force, Lorentz-Minkowsky's Metric, Universe Line, Energy

During **1905**, called his *annus mirabilis*, Einstein hypothesized that all physical laws must be the same in every inertial frame of reference and that the speed of light was always the same, regardless of the speed of the inertial system in which it is observed.

With this Einstein laid the foundations of the special theory of relativity.

Of course, this involved a new definition of both space and time variables, no longer absolute, but relative to the observer who measures them (see equations 5.6 and 5.7) and a representation in a space that is no longer three-dimensional but *four-dimensional*, in which time represents the fourth dimension.

Just as in the classical vision of space its three components (or dimensions) are equivalent and homogeneous with each other, the relativistic vision also assimilates the temporal dimension to the three spatial ones, making it perceptible in a different way by observers in different conditions.

8.1. Four-Position

Spacetime or *chronotope*[1] cannot be easily represented. To understand spacetime it is possible to imagine a flat Universe in two dimensions (x, y) as a surface (plane). The third Cartesian axis (z) is associated with time (t).

All the points of spacetime are called *events* and each of them corresponds to a phenomenon that is in a certain spatial position and at a certain time. Thus, each event is identified by four coordinates (x, y, z, t), one of which is time.

Referring to Figure 8.1 it is therefore possible to see one of the possible ways of representing a spacetime graph of a body that moves with uniform rectilinear motion. Another more intuitive way of representing spacetime is shown below in Figurue 8.2:

In special relativity, the four-vector, represented by a quadruple of values, is a vector of the Minkowski spacetime.

The four-position, corresponding to a point or *event* in the spacetime $\mathbb{R}^4$, will be therefore defined as follows:

$$\vec{\Sigma} := \left(t; \frac{x}{c}, \frac{y}{c}, \frac{z}{c}\right).$$

For sake of simplicity it is useful to define $\vec{\Sigma}$ only for the coordinates t and x. This is possible because, as demonstrated by the Lorentz transformations, given a rectilinear motion of any body along the x axis, the other two

---
[1]This term was introduced by H. Minkowski in 1908 to highlight the close link between space and time, established by the theory of relativity.

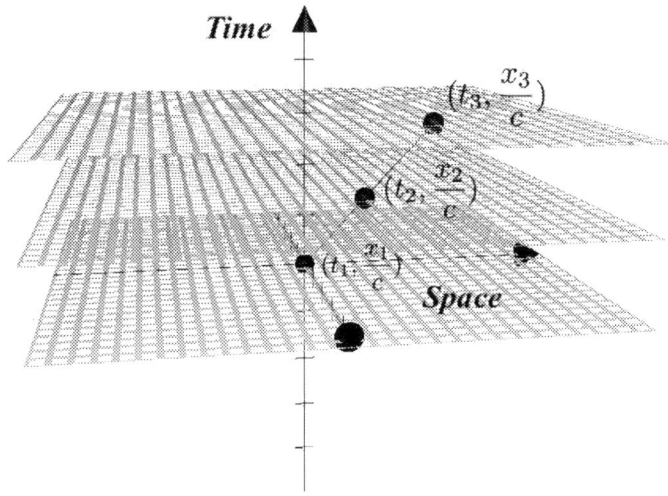

Figure 8.1.

coordinates will remain the same in any inertial reference frame:

$$\vec{\Sigma} := \left(t, \frac{\vec{x}}{c}\right). \tag{8.1}$$

8.2. Four-Velocity

The coordinates x, y and z denote proper length (with sign) while t is a non-proper time. The four-velocity is defined as the partial derivative of the position $\vec{\Sigma}$ with respect to proper time. This position will be theoretically justified in section 8.5..

Assuming this fact, the four-velocity is obtained:

$$\frac{\partial \vec{\Sigma}}{\partial \tau} = \left(\frac{\partial t}{\partial \tau}, \frac{\partial \frac{\vec{x}}{c}}{\partial \tau}\right) = \left(\gamma, \frac{1}{c}\frac{\partial \vec{x}}{\partial t}\frac{\partial t}{\partial \tau}\right) = \left(\gamma, \frac{\gamma}{c}\vec{v}\right)$$

$$\vec{V} := \frac{\partial \vec{\Sigma}}{\partial \tau} = \left(\gamma, \frac{\gamma}{c}\vec{v}\right). \tag{8.2}$$

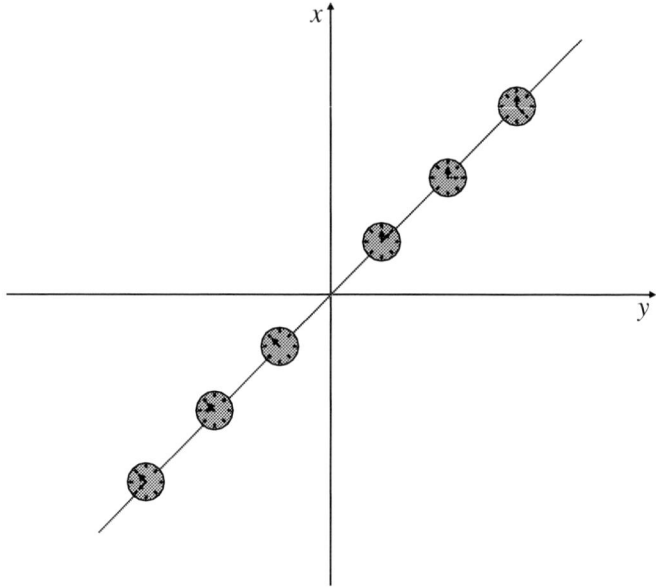

Figure 8.2.

Homogenization either with respect to times or spaces implies the usage of the so called *geometric units* in which, in particular *(i)* the speed is a *pure number* and *(ii)* $c = 1$ (as a matter of fact, $x/t = c \Leftrightarrow v = 1$, see below):

$$v \stackrel{t}{=} \frac{\frac{x}{c}}{t} \stackrel{s}{=} \frac{x}{ct}$$

where "t" and "s" above stands for homogenizing with respect to *time* and *space* respectively.

Note that by homogenizing with respect to *time* it results

$$\vec{V} = \left(\gamma, \frac{\gamma}{c}\vec{v}\right)$$

whose Lorentz-Minkowski norm is $|\vec{V}| = 1$. On the other hand, by homogenizing with respect to *space* it results

$$\vec{V} = (\gamma c, \gamma \vec{v})$$

so that $|\vec{V}| = c$. Without any homogenization, namely *by using natural units*, it results instead

$$\vec{V} = (\gamma, \gamma \vec{v})$$

and so $|\vec{V}| = \gamma\sqrt{1 - v^2}$.

8.3. Four-Acceleration

The acceleration corresponds to the partial derivative of the four-velocity $\vec{V}$ defined by 8.2 with respect to proper time. Therefore, the four-acceleration will be defined as follows:

$$\vec{A} := \frac{\partial \vec{V}}{\partial \tau}. \tag{8.3}$$

This definition will be deepened in the next chapter 9.

8.4. Lorentz-Minkowsky's Metric

In Euclidean geometry the modulus of Σ would be:

$$|\vec{\Sigma}| := \sqrt{t^2 + \left(\frac{x}{c}\right)^2}$$

But, according to the Lorentz-Minkowsky metric, it is instead:

$$|\vec{\Sigma}| := \sqrt{t^2 - \left(\frac{x}{c}\right)^2}$$

therefore:

$$|\vec{\Sigma}|^2 = t^2 - \left(\frac{x}{c}\right)^2.$$

This is possible in virtue of the so-called *West-Coast Convention*, according to which time is positive and space is negative.

8.5. Deriving with Respect to Proper Time

This section intends to theoretically justify the passages carried out in the previous sections of this chapter and the following chapter 9. Although short, it has a fundamental importance that should not be taken for granted.

In particular, it is necessary to justify the passage consisting of the derivative of $\vec{x}$ with respect to τ. In fact, in the previous chapters $\vec{x}$ was considered as a *proper length*, so:
$$\frac{d\vec{x}}{dt} = \vec{v}$$
therefore:
$$\frac{d\vec{x}}{d\tau} = \frac{d\vec{x}}{dt}\frac{dt}{d\tau} = \gamma\vec{v}.$$
In this way, it was possible to quote, as taught by Einstein himself in 1905, two quantities, *one proper* and the other *not proper*.

The questions that arise spontaneously are therefore three, here listed:

- Is it possible to quote two proper quantities?
- Is it possible to quote two non-proper quantities?
- Is it possible to derive $\vec{\Sigma}$ with respect to *a non proper time*?

The answers are presented in the sequel.

Quoting Two Proper Quantities

Considering two proper quantities, they must obviously be measured in the same reference system.

This simple statement definitively solves the problem of the section. In fact:

- If a **proper length** $\Delta\vec{\sigma}$ is measured then
$$\Delta\tau = 0$$
  in fact, the measured time in which to measure the proper length must necessarily be equal to zero, by definition of proper length.

- If a **proper time** $\Delta\tau$ is measured then
$$\Delta\vec{\sigma} = 0$$
  in fact, a null movement will be measured, as it is measured in the reference system in which the measurement of $\Delta\tau$ is taken, by definition of proper time itself.

Therefore, the quotient between two proper quantities has no physical sense and gives as a result 0 or *infinite*.

Quoting Two Non-Proper Quantities: the Celerity

The quotient between two non-proper quantities exists but does not have a precise physical meaning. As a matter of fact, in order to find a coherent result the experience must be in the same point at the same moment.

Anyway, it is at least theoretically admissible to give a significant both to the quotient of two non proper quantities and to the quotient of two suitable proper quantities, as herein detailed (where $\Delta\sigma_{\text{fix}}$ denotes the proper length measured by the fixed reference and the other notations go straightforward):

$$\frac{\Delta s}{\Delta t} = \frac{\Delta s_{\text{mov}}}{\Delta t_{\text{fix}}} = \frac{\Delta s_{\text{mov}}}{\gamma \Delta \tau_{\text{mov}}}$$
$$= \frac{\Delta \sigma_{\text{fix}}}{\gamma \Delta t_{\text{fix}}}$$

which is anyway equal to $\frac{v}{\gamma}$. Also

$$\frac{\Delta \sigma}{\Delta \tau} = \frac{\Delta \sigma_{\text{fix}}}{\Delta \tau_{\text{mov}}} = \frac{\gamma \Delta s_{\text{mov}}}{\Delta \tau_{\text{mov}}}$$
$$= \frac{\gamma \Delta \sigma_{\text{fix}}}{\Delta t_{\text{fix}}}$$

which is equal to γv. So

$$\frac{\Delta s}{\Delta t} = \gamma^2 \frac{\Delta \sigma}{\Delta \tau}.$$

that is

$$\frac{\Delta s_{\text{mov}}}{\Delta t_{\text{fix}}} = \gamma^2 \frac{\Delta \sigma_{\text{fix}}}{\Delta \tau_{\text{mov}}}.$$

The quotient $\frac{\Delta s_{\text{mov}}}{\Delta t_{\text{fix}}}$ is called **proper velocity**, or *celerity*.

Deriving the Four-Position with Respect to Proper Time

In this chapter four-velocity has been defined as follows (where, remember, $\vec{x}$ is a proper length):

$$\vec{V} := \frac{\partial \vec{\Sigma}}{\partial \tau} = \left(\frac{\partial t}{\partial \tau}, \frac{\partial \vec{x}}{\partial \tau} \right) = \left(\gamma, \frac{1}{c} \frac{\partial \vec{x}}{\partial t} \frac{\partial t}{\partial \tau} \right) = \left(\gamma, \frac{\gamma}{c} \vec{v} \right).$$

But, why is it necessary to derive the position vector with respect to proper time? Why is it not possible to derive it with respect to non-proper time?

The 4−position vector is an **absolute quantity**, which defines a point in $\mathbb{R}^4$. To derive an absolute quantity with respect to another quantity it is necessary that this other quantity is also absolute. This is because it is not possible to derive the 4−position with respect to something that is not absolute but is relative with respect to a further frame of reference.

Non-proper time is not an absolute quantity as it depends on the *proper time*.

8.6. Universe Lines

Minkowsky took Lorentz transformations and looked at them from a mathematical point of view. His idea of spacetime is presented below:

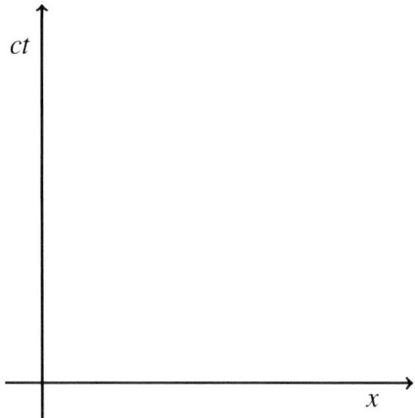

The question that can arise spontaneously is the following: why ct in the y axis? If a point P is defined by 4 variables, $P = (x, y, z, t)$, it has three units of length and one unit of time. To standardize them, the time is multiplied for the speed of light (always constant), obtaining a length.

The four-dimensional space $\mathbb{R}^4$ is simplified to $\mathbb{R}^2$ as the coordinates y and z are constant. The Lorentz equations written by using the factor $\beta := \frac{v}{c}$ are the following:

$$\begin{cases} x' = \gamma(x - \beta ct) \\ t' = \gamma\left(t - \beta\frac{x}{c}\right). \end{cases}$$

To find the vertical axis, it is set $ct' = 0$ which means $t' = 0$ and therefore $ct = \beta x$. This is the new axis of the Minkowsky plane. On the other hand, by putting $x' = 0$ so that $ct = \frac{x}{\beta}$, the perpendicular to the first-third quadrant is the line dividing the plane. Basically, the plane now becomes similar to this:

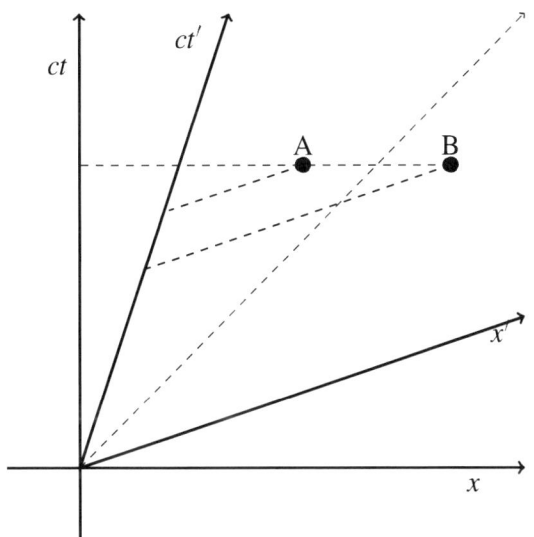

From this graph it is possible to derive the fundamental equation:

$$t_A = t_B \quad \text{but} \quad t'_A \neq t'_B.$$

8.7. A Look towards Dynamics

Energy

What is energy? Energy is an enigmatic concept, a mysterious phenomenon that is difficult to be defined. Some great minds including Richard Feynman tried to define energy. He said: *It is important to realize that in physics today, we have no knowledge of what energy is.* And, moreover, *there is a fact, or a law, governing all phenomena that are known to date... The law is called the conservation of energy. It states that there is a certain quantity, which we call energy that does not change in the manifold changes that nature undergoes. That is a most abstract idea, because it is a mathematical principle, it says that there is a numerical quantity which doesn't change when something happens. It is not a description of a mechanism, or anything concrete, it is just a strange fact that we can calculate some number and when we finish watching nature go through her tricks and calculate the number again..., it is the same.*

Even if energy is not known, it is possible to represent energy mathematically.

Total Energy

In the following it will be very useful to define $E := \gamma mc^2$. Here it is not taken into account the potential term of the energy.

Heat and Work

Although they are not rigorous, the following definitions are surely very vivid.

- *Work* is *travelling* energy due to the presence of a *force*.

- *Heat* is *travelling* energy generally due to the presence of a *difference of temperature*. In isothermal transformations a perfect gas receiving heat totally employs it for expansion into the external environment.

Four-Momentum

The four-momentum is merely defined to be as the product between mass and four-velocity:

$$\vec{P} := m\vec{V} = \left(\gamma m, \frac{\gamma}{c} m\vec{v}\right) \tag{8.4}$$

so its Lorentz-Minkowsky squared norm is

$$|\vec{P}|^2 := (\gamma m)^2 - \left(\frac{\gamma}{c} m\vec{v}\right)^2 = \gamma^2 m^2 - \frac{\gamma^2}{c^2} m^2 v^2 = m^2. \tag{8.5}$$

Thus, the norm of the 4−momentum is together with mass, c and the relative velocity another *invariant*.

Multiplying equation 8.5 by c^4, standing the fact[2] that the scalar momentum $\vec{p}$ equals to $\gamma m \vec{v}$, it easily follows that

$$\gamma^2 m^2 c^4 - \gamma^2 m^2 v^2 c^2 = m^2 c^4 \Leftrightarrow E^2 = c^4 m^2 + c^2 p^2 \quad (8.6)$$

where the total energy E of a massive body has been defined as $\gamma m c^2$. As a matter of fact this is a well posed definition standing that the following relationship holds

$$d\left(\gamma m c^2\right) = \frac{d\vec{p}}{dt} d\vec{x} \quad (8.7)$$

where - as mentioned above - for sake of simplicity it is assumed to have a body moving only along the x axis. In other words by posing $E := \gamma m c^2$ it's possible to resume the relativistic version of the so called *vis vivae* theorem.

At last, equation 8.7 follows by differentiating (8.6), using the definitions of $\vec{p}$ and E themselves. Indeed:

$$dE^2 = d\left(c^4 m^2 + c^2 p^2\right) \Leftrightarrow \boxed{E\,dE = c^2 p\,dp}$$

on the other hand

$$\begin{cases} E = \gamma m c^2 \\ \vec{p} = \gamma m \vec{v} \end{cases} \Rightarrow \frac{\vec{p}}{E} = \frac{\vec{v}}{c^2} \Rightarrow \boxed{\frac{p}{E} = \frac{v}{c^2}}$$

so

$$E\,dE = c^2 p\,dp = c^2 \frac{v}{c^2} E\,dp \Rightarrow dE = v\,dp = vF\,dt = F\,dx$$

as desired.

Let's define

$$\vec{f} := \frac{\partial \vec{p}}{\partial t}.$$

Whatever $\vec{f}$ will be it is possible to express $\vec{f}$ as $\vec{f} = \begin{pmatrix} f_\| \\ f_\perp \end{pmatrix}$ where $f_\|$ and $f_\perp$ are intended to be the components of $\vec{f}$ respectively parallel and perpendicular to $\vec{v}$.

---

[2]It's possible (even if not simple) to proof that both *the correspondence principle* and *the requirement for the momentum to be preserved in collisions* determine the equation $\vec{p} = \gamma m \vec{v}$.

Let's denote with $\hat{t}$ and $\hat{n}$ the pinors of $\vec{v}$ and $\vec{v}_\perp$ respectively. Standing the notations above it results:

$$f_\parallel = \frac{\partial \vec{p}_\parallel}{\partial t} = \frac{\partial (\gamma m \vec{v})_\parallel}{\partial t} = m\left(\frac{\gamma^3}{c^2}\vec{v}\cdot\vec{a}_\parallel \vec{v} + \gamma \vec{a}_\parallel\right) = m\left(\frac{\gamma^3}{c^2}v^2 a_\parallel \hat{t} + \gamma a_\parallel \hat{t}\right)$$

$$= m\left(\frac{\gamma^3}{c^2}v^2 \frac{\partial v}{\partial t}\hat{t} + \gamma \frac{\partial v}{\partial t}\hat{t}\right)$$

$$= m\gamma\left(\gamma^2 \frac{v^2}{c^2} + 1\right)\frac{\partial v}{\partial t}\hat{t}$$

$$= m\gamma^3 \frac{\partial v}{\partial t}\hat{t} = m\frac{\partial (\gamma v)}{\partial t}\hat{t}$$

$$= m\gamma^3 a_\parallel \hat{t}$$

and

$$f_\perp = \frac{\partial \vec{p}_\perp}{\partial t} = \frac{\partial (\gamma m \vec{v})_\perp}{\partial t} = m\gamma a_\perp \hat{n} = m\gamma \frac{v^2}{R}\hat{n}$$

where R denotes here the radius of curvature of the trajectory at the instant t.

Hence

$$\vec{f} = m\gamma^3 a_\parallel \hat{t} + m\gamma \frac{v^2}{R}\hat{n} = \begin{pmatrix} m\gamma^3 a_\parallel \\ m\gamma \frac{v^2}{R} \end{pmatrix} = \begin{pmatrix} m\gamma^3 \frac{\partial v_\parallel}{\partial t} \\ m\gamma \frac{v^2}{R} \end{pmatrix} = m\gamma \begin{pmatrix} \gamma^2 \frac{\partial v_\parallel}{\partial t} \\ \frac{v^2}{R} \end{pmatrix} = m\gamma \begin{pmatrix} \gamma^2 a_\parallel \\ a_\perp \end{pmatrix}$$

so, in particular, the relativistic force is not parallel to the acceleration.

Four-Force

Remember the notation already introduced:

- 4−velocity: $\vec{V} = \left(\gamma, \frac{\gamma}{c}\vec{v}\right)$
- 4−momentum: $\vec{P} = m\vec{V}$
- 4−acceleration: $\vec{A} = \frac{\partial}{\partial \tau}\vec{V}$.

Now define the 4−force as $\vec{F} := \frac{\partial \vec{P}}{\partial \tau}$. Standing these premises the following equation stands:

$$\vec{F} = m\vec{A},$$

in fact $\vec{F} = \frac{\partial (m\vec{V})}{\partial \tau} = m\frac{\partial \vec{V}}{\partial \tau} = m\vec{A}$, as in classical Newtonian mechanics.

Lorentz-Minkowsky's Spacetime 115

8.8. Some Interesting Solved Exercises

The following are some useful exercises for understanding and deepening the applications of special relativity.

Exercise 1. *(Gamma Factor)* *The γ Lorentz factor is always greater than or equal to one and equals one if and only if $v = 0$.*

Solution: Starting from the definition of γ,

$$\gamma = \gamma(v) = \frac{1}{\sqrt{1 - \left(\frac{v}{c}\right)^2}} \geq 1 \iff \frac{c - \sqrt{c^2 - v^2}}{\sqrt{c^2 - v^2}} \geq 0$$

which is equivalent to say

$$\sqrt{c^2 - v^2} \leq c \iff v^2 \geq 0$$

which is - obviously - always true. In particular, from the above mentioned equivalences it follows that $v = 0$ if and only if $\gamma = 1$.

Exercise 2. *(Dilating Times)* *Suppose you want to stretch the remaining time of your life of the 15% of it.*

(a) Calculate the speed necessary to obtain such a result.

(b) How fast should a rocket go if a 50% time dilation is desired?

(c) What about 99% dilation?

Solution: (a) It is important to remember the time dilation equation:

$$\Delta t = \gamma \Delta \tau$$

where $\gamma = \frac{1}{\sqrt{1 - \left(\frac{v}{c}\right)^2}}$ is the well known Lorentz factor. Therefore:

$$\Delta t = \Delta \tau + \frac{15}{100}\Delta \tau = \frac{115}{100}\Delta \tau$$

where

$$\gamma = \frac{1}{\sqrt{1 - \left(\frac{v}{c}\right)^2}} = \frac{115}{100} \iff 1.15^2 = \frac{1}{1 - \left(\frac{v}{c}\right)^2} \iff 1 - \left(\frac{v}{c}\right)^2 \simeq 0.76.$$

which implies $v \simeq 0.49\,c$.

(b)
$$\frac{1}{\sqrt{1-\left(\frac{v}{c}\right)^2}} = \frac{150}{100} \iff 2.25^2 = \frac{1}{1-\left(\frac{v}{c}\right)^2} \iff \left(\frac{v}{c}\right)^2 \simeq 0.56 \iff v \simeq 0.75\,c.$$

(c)
$$\frac{199}{100} = \frac{1}{\sqrt{1-\left(\frac{v}{c}\right)^2}} \iff \left(\frac{v}{c}\right)^2 \simeq 0.497 \iff v \simeq 0.86\,c.$$

Exercise 3. *(Twins with different ages)* Two twins live on planet Earth when one of them decides to undertake an interstellar journey getting on a spaceship that moves at 99% of the speed of light c.

(a) Knowing that the time spent in the reference of the spaceship (compared to the wristwatch of the first twin) is equal to 12 years, compute the time spent in the reference of the second twin (compared to his wristwatch).

(b) What about if the speed of the spaceship equals 30% c?

Solution: (a) It is necessary to exploit the equation of time dilation, from which given $\Delta\tau = 12\,\text{y}$:

$$\Delta t = \frac{1}{\sqrt{1-\left(\frac{v}{c}\right)^2}} \cdot \Delta\tau = \frac{1}{\sqrt{1-\left(\frac{0.99c}{c}\right)^2}} \cdot 12\,\text{y} = \frac{100}{\sqrt{199}} \cdot 12\,\text{y} \simeq 85\,\text{y}.$$

(b)
$$\Delta t = \frac{1}{\sqrt{1-\left(\frac{v}{c}\right)^2}} \cdot \Delta\tau = \frac{1}{\sqrt{1-\left(\frac{0.3c}{c}\right)^2}} \cdot 12\,\text{y} = \frac{10}{\sqrt{91}} \cdot 12\,\text{y} \simeq 13\,\text{y}.$$

Exercise 4. *(Interplanetary communications)* Consider two planets A and B whose proper distance is 2.4 ly, as depicted in Figure 8.3. An A inhabitant "α" decides to undertake a space trip to reach planet B while another inhabitant "β" remains on the planet A; the speed of α's spaceship is $\frac{c}{2}$. The moment α's spaceship touches planet B a light signal is sent towards planet A, as represented in Figure 8.4.

Lorentz-Minkowsky's Spacetime

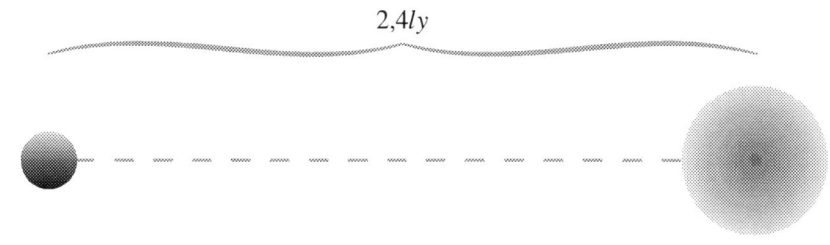

Figure 8.3.

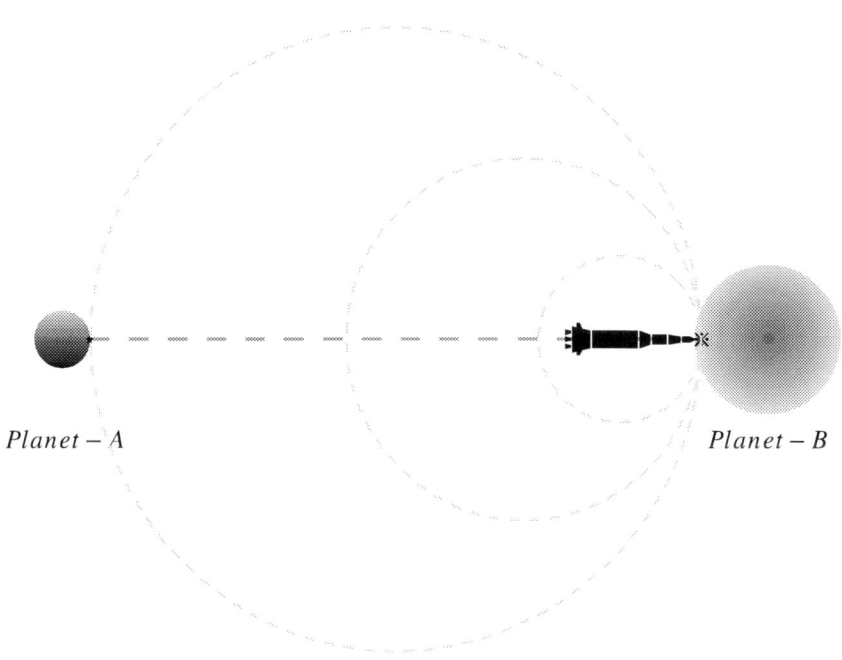

Figure 8.4.

(a) Determine how long the spaceship takes to reach planet B starting from planet A, both in α's and in β's frame of reference.

(b) In β's frame of reference, compute the time elapsed from the departure to the arrival of the light signal on planet A.

(c) Suppose now that in α's reference frame the time necessary for the spaceship to reach planet B equals 10 years. Under this hypothesis, what is its speed?

Solution: (a) The solution is immediate: it is necessary to exploit the equation of time dilation. In β's reference system, according to α's on board watch, the elapsed time will be equal to

$$\Delta t = \frac{2.4 \, \text{yc}}{\frac{c}{2}} = 4.8 \, \text{y}. \tag{8.8}$$

This makes it easy to get the time spent in α's reference:

$$\Delta \tau = \frac{\Delta t}{\gamma} = \Delta t \sqrt{1 - \left(\frac{v}{c}\right)^2} = 4.8 \sqrt{1 - \left(\frac{\frac{c}{2}}{c}\right)^2} \simeq 4.157 \, \text{y}.$$

(b) To solve this point it is simply necessary to add the result of the previous calculation 8.8 the time taken by the light to travel the distance of 2.4 light years, which is 2.4 years:

$$\Delta t = 4.8 + 2.4 = 7.2 \, \text{y}.$$

(c) Given that $\Delta \tau = 10$ y and $\Delta \sigma = 2.4$ yc the speed will be equal to

$$v = \frac{\frac{\Delta \sigma}{\gamma}}{\Delta \tau} \iff \gamma v = \frac{\Delta \sigma}{\Delta \tau}$$

which gives after straight calculation:

$$v = \frac{0.24c}{\sqrt{1 + 0.24^2}} \simeq 0.23c.$$

Exercise 5. *(Speed composition between relativistic racing cars)* There are two racing cars which move at relativistic speeds. The first one, say α, moves with respect to the reference integral to the ground with a speed of $0.8c$ while the second, β, moves with respect to the same reference with the speed of $0.6c$, in opposite direction, as shown by Figure 8.5.

Knowing that the proper length of β (namely the length measured in the reference system where both ends of the car are measured simultaneously) is 4.5 m, compute the non-proper length of β in α's reference system.

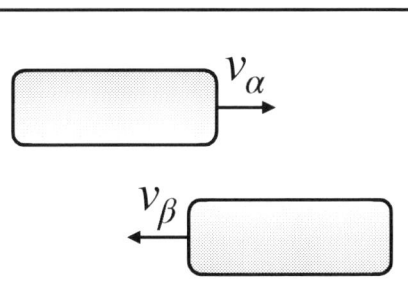

Figure 8.5.

Solution: It is necessary to refer to the reference of the α car, as the aim is to determine the length of β in the α car reference. Therefore it will be found the speed of β with respect to α. To do so it will be enough to exploit the relativistic composition of speed:

$$v_{\beta \to \alpha} = v_{\beta \to g} \oplus v_{g \to \alpha} = \frac{v_{\beta \to g} + v_{g \to \alpha}}{1 + \frac{v_{\beta \to g} \cdot v_{g \to \alpha}}{c^2}}$$

where, here, g, stands for the reference integral to the *ground*. Therefore:

$$v_{\beta \to \alpha} = \frac{-0.6c - (+0.8c)}{1 + \frac{-0.6c \cdot (-0.8c)}{c^2}} \simeq -0.95\,c.$$

Now it is possible to find the length of β in the α reference:

$$\Delta s_\beta = \frac{\Delta \sigma_\beta}{\gamma} = 4.5 \cdot \sqrt{1 - \left(\frac{-0.95c}{c}\right)^2} \simeq 1.46\,\text{m}.$$

Exercise 6. *(Relativistic Chase) Let's consider two particles, A and B moving along a same straight line in a certain laboratory. Particle A is running after B, its speed with respect to the laboratory is $v_a = \frac{4}{5}c$. On the other hand B is running away at speed $v_b = \frac{3}{5}c$. The initial distance between the two particles measured in the frame of the laboratory amounts to $d = 100$ km. Determine how much time does it take for the particle A to reach B*

(a) in the frame of the laboratory;

(b) in the frame of B;

(c) in the frame of A.

Solution: (a) In the frame of the laboratory (denoted in the following by L), fixed a coordinate system with its origin coincident with the initial position of A, the laws of motion of both A and B are the following: $s_A(\tau_L) = v_{A \to L} \cdot \tau_L$, on the other hand $s_B(\tau_L) = v_{B \to L} \cdot \tau_L + d$ so $s_A(\tau_L) = s_B(\tau_L)$ if and only if $\tau_L = d/(v_{A \to L} - v_{B \to L})$:

$$\tau_L = \frac{d}{v_{A \to L} - v_{B \to L}} = \frac{10^5}{\frac{4}{5}c - \frac{3}{5}c} = 5 \cdot 10^5 \, \text{lm} \simeq \frac{5}{3} \cdot 10^{-3} \, \text{s}$$

where, remember, $1 \, \text{lm} \simeq \frac{1}{3 \cdot 10^8} \, \text{s}$ is the time necessary for the light to walk in the vacuum along a straight path whose length is $1 \, \text{m}$.

(b) In the frame of reference coincident with the particle B, fix a system of coordinates positively oriented from A to B with its origin in B and denote by $\Delta \tau_A$ the time measured from A's wristwatch necessary to get to B:

$$|v_{A \to B}| = \frac{\frac{d}{\gamma(v_{A \to L})}}{\Delta \tau_A} \implies \Delta \tau_A = \frac{\frac{d}{\gamma(v_{A \to L})}}{|v_{A \to B}|} = \frac{\frac{10^5}{\frac{5}{3}}}{\frac{5}{13}c} = \frac{39}{25} \cdot 10^5 \, \text{lm}. \quad (8.9)$$

In other words, in the frame of B, $s_A(\tau_A) = v_{A \to B} \cdot \tau_A - \frac{d}{\gamma(v_{A \to L})}$ (the minus sign is necessary since the frame is oriented from A to B with its origin in B), on the other hand merely $s_B \equiv 0$ so that A overcome B if and only if $v_{A \to B} \cdot \tau_A = \frac{d}{\gamma(v_{A \to L})}$, which imply (8.9).

Note that **it is a nonsense** to contract d (which is, remember, the distance measured in the frame of the *laboratory*) by using $\gamma(v_{A \to B})$, since, in order to get to B it is needed to pass throw the *laboratory* frame. Ultimately, since d is measured with respect to the laboratory no other contraction of d is permitted rather than $d/\gamma(v_{A \to L})$ or $d/\gamma(v_{B \to L})$.

(c) Conversely, in the frame of reference coincident with the particle A, fix a system of coordinates positively oriented from A to B with its origin in A:

- in order to reach A, B must cover the non-proper distance *measured by B*: $\frac{d}{\gamma(v_{B \to L})}$.

Lorentz-Minkowsky's Spacetime   121

- As usual, let's denote by $\Delta\tau_B$ the interval of proper time beaten by B's clock necessary to reach A.

- Since A is moving itself with respect to the laboratory (denoted in the following by L) the quotient between the two above mentioned expression is not $v_{A \to L}$ but $v_{B \to A}$.

So

$$|v_{B \to A}| = \frac{\frac{d}{\gamma(v_{B \to L})}}{\Delta\tau_B} \implies \Delta\tau_B = \frac{\frac{d}{\gamma(v_{B \to L})}}{|v_{B \to A}|} = \frac{\frac{10^5}{\frac{5}{4}}}{\left|-\frac{5}{13}c\right|} = \frac{52}{25} \cdot 10^5 \text{ lm}. \quad (8.10)$$

In other words, in the frame of A, $s_B(\tau_B) = v_{B \to A} \cdot \tau_B + \frac{d}{\gamma(v_{B \to L})}$, on the other hand merely $s_A \equiv O$ so that A overcome B if and only if $v_{B \to A} \cdot \tau_B = -\frac{d}{\gamma(v_{B \to L})}$, which imply (8.10).

Exercise 7. *(De Sitter Paradox)* Consider a star, say α, revolving around a black hole or around a more massive star β. Star α revolves around its companion β in a certain time, namely the period T. Very far away, at distance d, in the same orbital plane there is planet Earth. Now suppose to contradict the second postulate of special relativity on the constancy of the speed of light. So, according to Galileo's and Newton's mechanics, the speed of light is no longer constant and it is possible to sum it with other velocities.

Should it be possible that light starting in A will arrive to Earth in the same instant of the analogous light starting in B? Referring to Figure 8.6 and assuming classical mechanics give condition under which light starting in A arrives simultaneously with respect to the analogous starting in B.

Solution: Referring to Figure 8.6, consider the position A: α is now rotating away in opposite direction with respect to Earth. How long does the light take to reach Earth? According to classical mechanics it would take $t_A = \frac{d}{c-v}$. On the other hand, standing the fact that the period of α around β is T, α takes $T/2$ to go from position A to position B. In such a position, α moves toward Earth with speed v: here the velocities, c and v sum each other, so, according to Galileo, the time needed for the light to arrive to

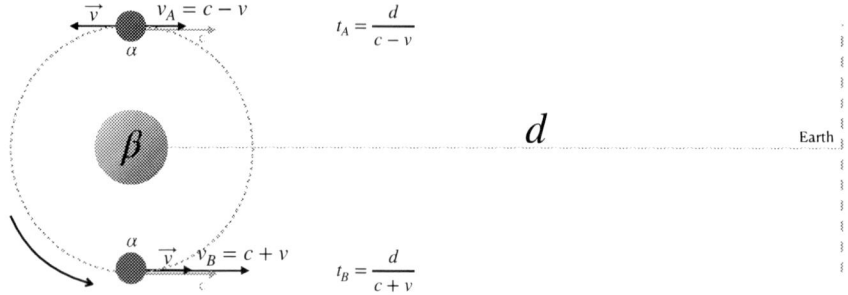

Figure 8.6.

planet Earth is $t_B = \frac{d}{c+v}$. Standing the above it is obvious that light starting from A travels slower than light starting from B, namely $t_A > t_B$. Should it be possible that light starting in A will arrive to Earth in the same instant with respect to light starting in B? Standing the fact that - assuming classical mechanics - light A is slower than light B, yes, it is surely possible, in fact it suffices that

$$t_A = t_B + \frac{T}{2} \iff \frac{2dv}{c^2 - v^2} = \frac{T}{2}. \tag{8.11}$$

Among the numerous revolving stars there are lots of them which satisfy relation 8.11, so that all the stars with suitable values of d and v should appear doubled.

On the other hand, as observed by De Sitter, there is no case in which a star satisfying equation 8.11 double itself.

The unique way to solve this paradox is to assume the initial hypothesis to be false. So, as the relativistic theory predicts, the speed of light is independent on the motion of the source and always $t_A = t_B$. So t_A is always less than $t_B + \frac{T}{2}$ and consequentially also mathematically no star can be doubled, theory agrees with experience and the paradox is solved.

Exercise 8. *(Relativistic Cannon) A special cannon fires particles at the incredible speed of 75% c. Such a cannon is mounted on the top of a spaceship whose speed with respect to the fixed stars amounts to $|\vec{v}| = \frac{1}{4}c$, as exemplified in Figure 8.7. Determine:*

(a) the speed of the particles relative to the spaceship;

(b) *if the particles were photons, how fast would they move relative to the spaceship?*

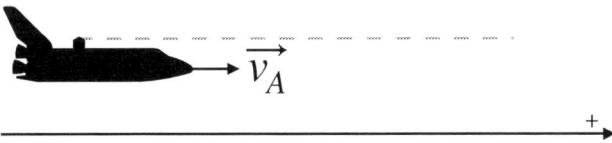

Figure 8.7.

Solution: (a) Given "p" the particles, "a" the spaceship and "s" the fixed stars:

$$v_{p \to a} = \frac{v_{p \to s} + v_{s \to a}}{1 + \frac{v_{p \to s} \cdot v_{s \to a}}{c^2}} = \frac{v_{p \to s} - v_{a \to s}}{1 - \frac{v_{p \to s} \cdot v_{a \to s}}{c^2}} = \frac{\frac{3}{4}c - \frac{1}{4}c}{1 - \frac{\frac{3}{4}c \cdot \frac{1}{4}c}{c^2}} = \frac{8}{13}c.$$

(b) Just remember the second postulate of special relativity: *the speed of light is the same in all inertial reference frames*, that is, in the references where Newton's first law holds. So the speed of the photons with respect to the spaceship will be exactly equal to $c = 299\,792\,458\,\frac{m}{s}$.

Exercise 9. *(Ultra fast-Service Station) A rocket travels at speed $v = 0.60c$ and flies by a space station in which a device detects its passage. As the rocket's tail passes in front of the device, it emits a flash of light in the same direction of the rocket's motion. The length of the rocket, in the reference system integral with it is $L = 150\,\text{m}$.*

(a) *In the reference system integral with the rocket, how long does it take light to reach the front of the rocket?*

(b) *In the reference system integral with the space station, how long does it take light to reach the front of the rocket?*

(c) *In the reference system integral with the space station how far from the station does the light beam reach the front of the rocket?*

Solution - first method: As stated in 6.3., the equations for *space contraction* and *time dilation* describe physical models with *two* frame of reference. On the other hand Lorentz equations describe the motions of physical models with *three* frame of reference. In this exercise there are actually three frames, namely *the space station, the rocket* and the *ray of light*. However the solution goes straight anyway without using Lorentz equations approaching the problem as in exercise 6.

(a) In the reference system integral with the rocket, by the second postulate of special relativity the speed of light with respect to the rocket is always c. On the other hand, with respect to the rocket its length is L so that the time needed to the light beam to reach the front of the rocket is:

$$\frac{L}{c} \simeq \frac{150}{3 \cdot 10^8} = 5 \cdot 10^{-7} \text{ s}.$$

(b) In the frame of the space station fix a one dimensional coordinate system positively oriented towards the direction of motion of the rocket, with its origin coincident with the device. Under these assumption the laws of motion of both the *front of the rocket* and the *ray of light* are respectively the following:

$$s_{\text{rocket}}(t) = v \cdot t + \frac{L}{\gamma(v)}$$

and

$$s_{\text{light}}(t) = c \cdot t$$

so the ray light reaches the front of the rocket if and only if

$$s_r(t) = s_l(t) \iff v \cdot t + \frac{L}{\gamma(v)} = c \cdot t \iff t = \frac{L}{(c-v) \cdot \gamma(v)} = \frac{150}{\frac{4}{10}c \cdot \frac{5}{4}} \simeq 10^{-6} \text{ s}$$

where $\gamma(v) = \gamma(\frac{3}{5}c) = \frac{5}{4}$. **(c)** So the ray of light reaches the front of the rocket $s_l \left(10^{-6}\right)$ meters far from the device of the space station:

$$s_l\left(10^{-6}\right) = c \cdot 10^{-6} \simeq 300 \text{ m}.$$

Note that, remembering what previously stated *"in this exercise there are actually three frames"*, given this particularity **it would constitute an evident big mistake** to answer (a) using the result (b) by invoking time dilation equation $\frac{\Delta t}{\gamma} = \frac{10^{-6}}{\frac{5}{4}}$.

Lorentz-Minkowsky's Spacetime

Solution - second method: First of all fix two coordinate systems, the first one integral with the device, say x, the second, x', integral with the spaceship, with both their origins coincident with the tail of the spaceship at the instant $t_0 = t'_0$, as shown in Figure 8.8.

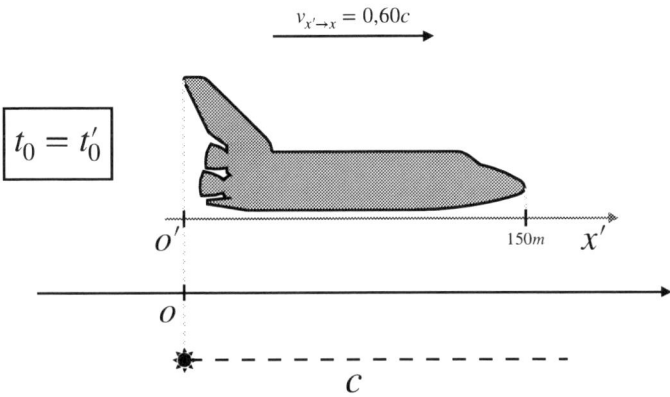

Figure 8.8.

In order to give an alternative solution the following Lorentz transformation equations will be employed:

$$\begin{cases} \Delta t' = \gamma\left(\Delta t - \frac{v_{x' \to x}}{c^2}\Delta x\right) \\ \Delta x' = \gamma(\Delta x - v_{x' \to x}\Delta t) \end{cases} \iff \begin{cases} \Delta t = \gamma\left(\Delta t' + \frac{v_{x' \to x}}{c^2}\Delta x'\right) \\ \Delta x = \gamma(\Delta x' + v_{x' \to x}\Delta t'). \end{cases}$$

(a) Regarding this point it is the same as that already exploited in the first method: $\Delta t' = 5 \cdot 10^{-6}$ s. For the other two points it is sufficient to exploit the above mentioned Lorentz transformations equations. In detail:

(b)

$$\Delta t = \gamma\left(\Delta t' + \frac{v_{x' \to x}}{c^2}\Delta x'\right) = \frac{5}{4}\left(5 \cdot 10^{-7} + \frac{0{,}6c}{c^2} \cdot 150\right) \simeq 10^{-6} \text{ s}.$$

(c)

$$\Delta x = \gamma\left(\Delta x' + v_{x' \to x}\Delta t'\right) = \frac{5}{4}\left(150 + 0{,}6c \cdot 5 \cdot 10^{-7}\right) \simeq 300\,\text{m}.$$

Exercise 10. *(Fast Service Station) The same situation prospected in exercise 9 with the difference that now the space station delivers the rocket a body whose speed is* $0.80c$ *(with respect to the space station).*

Solution - first method: (a) The solution is similar to that of the previous exercise. In detail, in the reference system integral with the rocket, by the theorem of composition of velocities the speed of the body with respect to the rocket is:

$$v_{\underline{body} \to \underline{rocket}} = \frac{v_{b \to station} + v_{s \to r}}{1 + \frac{v_{b \to s} \cdot v_{s \to r}}{c^2}} = \frac{v_{b \to station} - v_{r \to s}}{1 - \frac{v_{b \to s} \cdot v_{r \to s}}{c^2}} = \frac{5}{13}c.$$

Standing the fact that the rocket's length with respect to itself is L, the time needed to the light beam to reach the front of the rocket is:

$$\frac{L}{v_{b \to r}} = \frac{150}{\frac{5}{13}c} \simeq 1{,}3 \cdot 10^{-6} \text{ s}.$$

(b) Also this time, under the same assumption of the analogous previous solution it results:

$$s_{r \to s}(t) = v_{r \to s} \cdot t + \frac{L}{\gamma(v_{r \to s})}$$

where $\gamma(v_{r \to s}) = \gamma(v) = \frac{5}{4}$, and

$$s_{b \to s}(t) = v_{b \to s} \cdot t$$

so the body reaches the front of the rocket if and only if

$$s_{r \to s}(t) = s_{b \to s}(t) \iff v_{r \to s} \cdot t + \frac{L}{\gamma(v_{r \to s})} = v_{b \to s} \cdot t$$

$$\iff t = \frac{L}{(v_{b \to s} - v_{r \to s})\gamma(v_{r \to s})} = \frac{150}{\frac{1}{5}c \cdot \frac{5}{4}} \simeq 2 \cdot 10^{-6} \text{ s}.$$

(c) The ray of light reaches the front of the rocket $s_{b \to s}(2 \cdot 10^{-6})$ meters far from the device of the space station:

$$s_{b \to s}(2 \cdot 10^{-6}) = \frac{4}{5}c \cdot 2 \cdot 10^{-6} \simeq 480 \, \text{m}.$$

Solution - second method: (a) Regarding this first part it is the same as that already exploited in the first method: $\Delta t' = 1{,}3 \cdot 10^{-6}$ s.

For the other two points, by using the same notations previously introduced in the analogous solutions, it is sufficient to exploit the following Lorentz equations:

$$\begin{cases} \Delta t = \gamma(v_{x' \to x}) \left(\Delta t' + \frac{v_{x' \to x}}{c^2} \Delta x' \right) \\ \Delta x = \gamma(v_{x' \to x}) \left(\Delta x' + v_{x' \to x} \Delta t' \right) \end{cases}$$

where $\gamma(v_{x' \to x}) = \gamma(v_{r \to s}) = \frac{5}{4}$. As a matter of facts: **(b)**

$$\Delta t = \gamma \left(\Delta t' + \frac{v_{x' \to x}}{c^2} \Delta x' \right) = \frac{5}{4} \left(1{,}3 \cdot 10^{-6} + \frac{0{,}6c}{c^2} \cdot 150 \right) \simeq 2 \cdot 10^{-6} \text{ s}.$$

(c)

$$\Delta x = \gamma \left(\Delta x' + v_{x' \to x} \Delta t' \right) = \frac{5}{4} \left(150 + 0{,}6c \cdot 1{,}3 \cdot 10^{-6} \right) \simeq 480 \text{ m}.$$

Exercise 11. *(A one way trip to the Pleiades) An astronaut, aboard his spaceship, decides to make a journey to the Pleiades, 400 light-years away from Earth, as shown in Figure 8.9. When he arrives at his destination, he looks at his on-board clock which shows an elapsed time of 5 years. Determine the elapsed time in the reference of an Earth inhabitant (relative to his wristwatch).*

Figure 8.9.

Solution - first method: First of all it is necessary to find the relative speed:

$$v = \frac{\Delta s}{\Delta \tau} = \frac{\Delta \sigma}{\gamma \cdot \Delta \tau} = \frac{\Delta \sigma}{\frac{1}{\sqrt{1-\left(\frac{v}{c}\right)^2}} \cdot \Delta \tau} = \frac{\Delta \sigma}{\Delta \tau} \cdot \sqrt{1 - \left(\frac{v}{c}\right)^2}$$

therefore:

$$v = \frac{400yc}{5y} \cdot \sqrt{1 - \frac{v^2}{c^2}} \quad \Leftrightarrow \quad v^2 = 6400 \cdot c^2 - 6400c^2 \cdot \frac{v^2}{c^2} \quad \Leftrightarrow \quad v^2 = \frac{6400}{6401} \cdot c^2.$$

Now it is possible to find the elapsed time in the Earth reference frame:

$$\Delta t = \gamma \cdot \Delta \tau = \frac{1}{\sqrt{1 - \frac{v^2}{c^2}}} \cdot \Delta \tau = \frac{1}{\sqrt{1 - \frac{6400}{6401}}} \cdot 5\,y \simeq 400.0312\,y.$$

Solution - second method: The solution of the problem with this second method is based on exploiting the *invariant interval*, as reported below:

$$(\Delta t)^2 - \left(\frac{\Delta x}{c}\right)^2 = (\Delta t')^2 - \left(\frac{\Delta x'}{c}\right)^2$$

which is

$$(\Delta \tau_1)^2 - \left(\frac{\Delta \sigma_1}{c}\right)^2 = (\Delta \tau_2)^2 - \left(\frac{\Delta \sigma_2}{c}\right)^2$$

where $\Delta \sigma_1$ is measured in the spaceship reference, and equals 0 since the reference with respect to itself does not undergo any spatial variations, while $\Delta \tau_1$ corresponds to the time elapsed in the spaceship reference. On the other hand, in the Earth reference frame $\Delta \sigma_2$ corresponds to the Earth-Pleiades *own distance*, while $\Delta \tau_2$ corresponds to the elapsed time in the reference of a terrestrial inhabitant. Therefore:

$$(5y)^2 - 0 = (\Delta \tau_2)^2 - \left(\frac{400\,yc}{c}\right)^2$$

so that

$$\Delta \tau_2 = \sqrt{25 + 16000}\,y \simeq 400.0312\,y.$$

Exercise 12. *(Fast Cookie Factory)* A biscuit factory uses a conveyor belt which moves at the amazing speed of $\frac{c}{3}$. The biscuit dough is placed on the conveyor belt and is hit by a punch which should make it assume a circular shape. Determine:

(a) the shape that the biscuits would take if the punch had a circular shape;

(b) the shape of the punch so that the biscuits appear circular.

Solution: **(a)** Since the conveyor belt moves at relativistic speeds, the shape of the biscuits cannot be the same as that obtained in a non-relativistic factory. In detail, in the reference of the factory, the biscuit dough undergoes a contraction in the same direction of motion, as established by the contraction of the distances. If S denotes the length of the cookie dough in the reference of the conveyor belt and S' the analogous one in the reference of the factory this means that

$$S' = \frac{S}{\gamma} = S\sqrt{1 - \left(\frac{\frac{c}{3}}{c}\right)^2} \simeq 0.943 \cdot S.$$

Now, it is important to remember that the dough undergoes a contraction only in the direction of motion, while it does not undergo any change in the direction perpendicular to it. For this reason, when the punch, whose shape doesn't undergo any variation, being it perpendicular to its motion, hits the biscuit dough, it gives a circular shape to the biscuit, which, however, is still on the conveyor belt. Once out of it, it will be lengthened by the factor γ along the diameter having the same direction of motion.

Its final shape is represented in Figure 8.10, where r is the radius of the circumference of the punch and the diameter $\overline{AB}$ is equal to

$$\overline{AB} = 2\gamma r = \frac{2r}{\sqrt{1 - \left(\frac{\frac{c}{3}}{c}\right)^2}} \simeq 2.12 \cdot r.$$

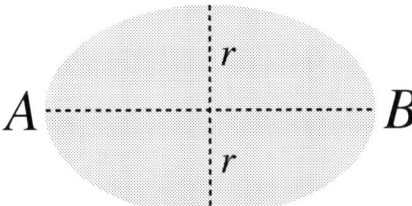

Figure 8.10.

(b) To ensure that the biscuits arrive in supermarkets circularly shaped, the diameter S of the punch in the same direction of the conveyor belt has to be contracted by the same γ factor:

$$S = \frac{S'}{\gamma} = S'\sqrt{1 - \left(\frac{\frac{c}{3}}{c}\right)^2} \simeq 0.943 \cdot S'.$$

Exercise 13. *(Downhill on a wagon) At a certain instant a wagon is moving along an inclined plane with speed v_0 directed along the plane. Suppose a ball is dropped inside the wagon at a certain height from the wagon's floor. Determine if the ball will fall perpendicular to the floor of the wagon or if it will touch the ground in front of or behind it.*

Solution: Referring to Figure 8.11, fix a canonical 2−dimensional Cartesian coordinate system integral to the moving wagon and consider the ball at height y_0 with respect to the wagon's floor. Now, let's describe all the forces acting on the ball and the wagon itself.

The ball, having mass m, is subject to the gravity force $m\vec{g}$, which can be split up into its two Cartesian components, namely $mg\sin\alpha$ and $mg\cos\alpha$. On the other hand, the wagon, whose mass is M, is subject to $M\vec{g}$ which corresponds to the two components $Mg\sin\alpha$ and $Mg\cos\alpha$. Moreover, the wagon is subjected to the constraint reaction of the plane, having opposite direction with respect to the vertical component of the gravity, so that these two vertical components cancels each other and the only horizontal component $Mg\sin\alpha$ remains.

At each instant $Mg\sin\alpha = Ma_M$ which imply $a_M = g\sin\alpha$, and $mg\sin\alpha = ma_m$ so that $a_m = g\sin\alpha = a_M$, that is, for every t, both the wagon and the ball are subject to the same acceleration $g\sin\alpha$. For this reason, considering the kinematics inside the wagon, by virtue of SEP$_E$ (see 4.5.) it is possible *to forget* the acceleration horizontal component acting both on the ball and on the wagon.

Therefore the only force acting on the ball remains the vertical component of the gravity, which is directed along the perpendicular line to the wagon's floor. So the ball must fall perpendicular to the floor as if the wagon were stationary on a horizontal plane.

Exercise 14. *Find the momentum of a proton whose velocity is*

$$\vec{v} = \begin{pmatrix} 343 \\ \frac{c}{10} \end{pmatrix} \frac{m}{s}$$

with respect to a certain coordinate frame.

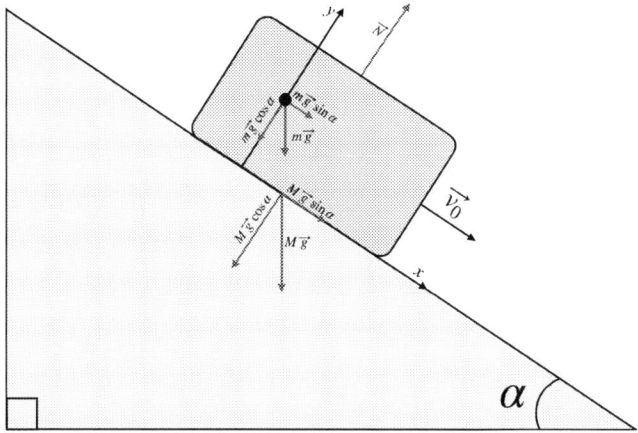

Figure 8.11.

Solution: The momentum of the proton is

$$|\vec{p}| = \gamma \cdot m \cdot |\vec{v}| = \frac{m \cdot v}{\sqrt{1 - \left(\frac{v}{c}\right)^2}}$$

where $m \simeq 1.6726 \cdot 10^{27}$ kg and the speed is approximately $\frac{c}{10} \simeq 3 \cdot 10^7$ m/s. Therefore, the requested momentum amounts to approximately

$$|\vec{p}| \simeq \frac{1.6726 \cdot 10^{-27} \cdot 3 \cdot 10^7}{\sqrt{1 - \left(\frac{3 \cdot 10^7}{3 \cdot 10^8}\right)^2}} \simeq 5.043 \times 10^{-20} \text{ kg} \cdot \frac{\text{m}}{\text{s}}$$

Exercise 15. (*A tan truck can solve all the world's problems*) *Consider a half-litre bottle of water. Determine (a) the total energy of this bottle. (b) Since the world's need for energy consists in approximately 1 ZJ (namely one "zetta joule" which is 10^{21} J.) calculate the amount of purified water sufficient to satisfy it.*

Solution: (a) Purified water density is $1000 \, \text{kg}/\text{m}^3$, so that the mass of the water inside the bottle equals to 0.5 kg. Therefore:

$$E = \gamma m c^2 = 0.5 \cdot 299\,792\,458^2 \simeq 4.5 \cdot 10^{16} \text{ J}$$

where γ is set to 1 since water is considered stationary. **(b)** The mass of stationary water corresponding to 1 ZJ satisfies the equation

$$E = mc^2 = 10^{21}\,\text{J}$$

so that

$$m = \frac{10^{21}}{c^2} \simeq 11\,111\,\text{kg}$$

correspondingly to approximate one tank truck, amazing!

Exercise 16. *(Relative motion between three particles)* *There are three particles: particle one moves with speed equal to $\frac{c}{2}$ with respect to particle two, which moves with speed equal to $\frac{c}{2}$ with respect to particle three. This particle moves itself with speed equal to $\frac{c}{2}$ with respect to the ground. Compute the speed of particle one with respect to the ground.*

Solution: It results

$$v_{1 \to g} = \frac{v_{1 \to 3} + v_{3 \to g}}{1 + \frac{v_{1 \to 3} \cdot v_{3 \to g}}{c^2}}$$

where

$$v_{1 \to 3} = \frac{v_{1 \to 2} + v_{2 \to 3}}{1 + \frac{v_{1 \to 2} \cdot v_{2 \to 3}}{c^2}} = \frac{\frac{c}{2} + \frac{c}{2}}{1 + \frac{\frac{c}{2} \cdot \frac{c}{2}}{c^2}} = \frac{4}{5}c.$$

Therefore:

$$v_{1 \to g} = \frac{\frac{4}{5}c + \frac{c}{2}}{1 + \frac{\frac{4}{5}c \cdot \frac{c}{2}}{c^2}} = \frac{13}{14}c.$$

Exercise 17. *A relativistic particle is accelerated so that its total energy is $6.45 \cdot 10^{-8}$ J and its momentum amounts to $21.3 \cdot 10^{-17}$ kg $\cdot$ m/s. Determine the particle's final speed.*

Solution: Since

$$E = \gamma \cdot m \cdot c^2$$

and

$$|\vec{p}| = \gamma \cdot m \cdot |\vec{v}|$$

by quoting them it results:

$$\frac{E}{p} = \frac{c^2}{v} \iff \frac{6.45 \cdot 10^{-8}}{21.3 \cdot 10^{-17}} = \frac{c^2}{v}$$

so that
$$v \simeq 2.97 \cdot 10^8 \, \frac{m}{s}.$$

Chapter 9
The Accelerated Motion

> *This chapter intends to fill the gap between the kinematic of uniform motion and general relativity, in order to underline that special relativity does not end with uniform motions. This chapter probably represents the true heart of the whole book and a small glimpse of a topic with thousand of possibilities and applications.*
>
> *Once 4−position and 4−velocity have been defined, it is possible to dedicate a discussion of the equations of motion for an accelerated motion. In particular, it is possible to compare the uniformly accelerated motion in classical mechanics with the relativistic one.*

Keywords: Accelerated motion, MITCF, Hyperbolic motion, Rindler's metric, Horizons, Cosmology, Universe's shape, Hubble's Law

The first point to consider is that what classically is a uniformly accelerated motion does not exist relativistically: in fact, the acceleration is not an invariant, so it will be measured as constant only in a reference system.

9.1. A Common Misconception about SR

Can special relativity handle acceleration? Yes, SR can handle accelerations[1] even though it is set up for inertial frames. As a matter of fact, instantaneous inertial frames can be defined so that an accelerating body moves in a continuous fashion from one instantaneous inertial frame[2] to another. In short, special relativity can be used for every type of acceleration **except for the whole gravity field** which pervades the entire spacetime. Note that SR is even suitable to handle *local uniform*[3] gravitational fields, since - by SEP - they can be emulated by a suitable anti-parallel uniform[4] relativistic acceleration determined by an electromagnetic, strong or weak force acting on the frame itself, without any difference affecting the results of the physical experiments. On the other hand, if, instead of *local*, the equivalence established by SEP was *global*, without any requirement about the *uniformity*, then it would originate an evident logical contradiction. As a matter of fact, there will be simultaneously (i) SR is not able to handle the gravity field pervading the entire Universe and (ii) since SR can surely handle any non-gravitational accelerated motion, being as per the absurd hypothesis gravitational fields equivalent to the effects of any force in nature (electromagnetic, strong and weak) SR would handle also gravitational field pervading the entire Universe, which is, obviously, a contradiction.

The equivalence established by SEP between the gravity field and other appropriate acceleration fields is true only *locally*, and the neighbourhood of any fixed point $P = (x^0, x^1, x^2, x^3)$ of the spacetime is as much as narrow as much the spacetime is curved in P. Besides, in such a neighbourhood the spacetime is shaped as the Lorentz-Minkowsky space, namely, it is flat, its curvature is identically zero, so the grid velocity is zero and no gravitational relativistic effect such as time contractions apply. In other words, *locally*, SR handles perfectly any accelerated, even gravitational, motion.

---

[1]Not necessarily uniform, see further for the precise definition.

[2]Such a frame is called *Momentarily Inertial Tangent Comoving (reference) Frame*, in brief, *MITCF*.

[3]Locality is a necessity that comes from the requested hypothesis of *uniformity* regarding the gravitational field.

[4]Here the uniformity has to be intended instant per instant, namely, for each fixed instant the field of accelerations has to be equal in each point so that in two different instant the accelerations are possibly different.

The Accelerated Motion 137

On the other hand, globally on the entire manifold - the spacetime, grid velocity is *not* zero, that is, points belonging to spacetime which are accelerated by gravity move both in space and in time along the coordinate grid, so that as much the gravity acceleration is, as much the spatial and temporal coordinates velocity increases[5]. In detail, as much the temporal grid velocity increases, as much the Lorentz factor γ increases, so that the proper time decreases realizing the above prospected *gravitational proper time contraction*. All these effects which are consequences of grid velocity are handled only by GR.

It is necessary to use at least one idea coming from the differential geometry. First of all take notice that gravity, that is the acceleration field g, pervades the whole spacetime. Besides, taking into account another of the so many simple brilliant Einstein's ideas, the distribution of mass-energy in spacetime determines its geometry, namely the metric tensor g. The metric tensor determines in turn the curvature of the spacetime and so the acceleration field g itself. Note that the *same letter*, g, is used to denote the *same phenomena*, from either the points of view, the physical one and the mathematical one.

Switching again to differential geometry it is not possible to define the metric tensor g on a unique chart that describes the whole Universe as a suitable open set of the Euclidean $\mathbb{R}^4$. As a matter of fact, the tensor g is defined over a collection of compatible charts each of the models a piece of spacetime as an open subset of $\mathbb{R}^4$, endowed with the Lorentz-Minkowsky metric tensor. In short, mass-energy determine the metric g, globally defined over spacetime, which determines itself the geometry of the Universe and therefore the acceleration field g which permeates the whole Universe.

Only locally (that is remaining *inside* each local chart) the tensor g is uniform so that the spacetime's curvature is zero and the grid velocity is zero itself (see further).

So, how is it possible that SR can handle all accelerations *except* the g field, being possible (accepting SEP) to emulate g by electromagnetic, strong or weak forces? As already underlined in this discussion, the problem lies in the need to reconcile local with global aspects: electromagnetic, strong and weak forces act only *locally* in spacetime, namely, they create

---

[5]In such a manner that the difference between their square remains always equal to c^2.

only localized acceleration fields, on the contrary, g is something that globally permeates the spacetime as a whole. Only locally g behaves as the acceleration field determined by the above-mentioned forces.

9.2. Definition of Uniformly Accelerated Motion

A body is said to be uniformly accelerated if its acceleration four-vector $\vec{A}$ has constant magnitude, say $\vec{A} \cdot \vec{A} = A^2 \geq 0$ and has constant spatial direction.

The definition of spatial acceleration $\vec{a}$ will be:

$$\vec{a} = \frac{d\vec{v}}{dt}. \tag{9.1}$$

The relationship between $\vec{A}$ and $\vec{a}$ will be deepened in 9.6.

9.3. Defining Four-Acceleration

The starting point is the definition of the 4−acceleration given in chapter 8 (see equation 8.3):

$$\vec{A} = \vec{A}(t, \sigma) := \frac{\partial \vec{V}}{\partial \tau}.$$

Now it is possible to start calculating $\vec{A} = \vec{A}(t, \sigma)$:

$$\vec{A} := \frac{\partial \vec{V}}{\partial \tau} = \left(\frac{\partial \gamma}{\partial \tau}, \frac{1}{c} \frac{\partial (\gamma \vec{v})}{\partial \tau} \right).$$

The Derivative of γ with respect to τ

First, the calculation of $\frac{\partial \gamma}{\partial \tau}$ is carried out:

$$\frac{\partial \gamma}{\partial \tau} = \left(\frac{\partial}{\partial t} \frac{\partial t}{\partial \tau} \right) \gamma = \gamma \frac{\partial \gamma}{\partial t}$$

It is now necessary to calculate $\frac{\partial \gamma}{\partial t}$:

$$\frac{\partial \gamma}{\partial t} = \frac{\partial \left(1 - \left(\frac{v}{c}\right)^2\right)^{-\frac{1}{2}}}{\partial t} = -\frac{1}{2} \left(1 - \left(\frac{v}{c}\right)^2\right)^{-\frac{3}{2}} \cdot \frac{\partial}{\partial t} \left(-\frac{\vec{v} \cdot \vec{v}}{c^2}\right) = \frac{1}{2} \frac{\gamma^3}{c^2} (\vec{a} \cdot \vec{v} + \vec{v} \cdot \vec{a}) = \frac{\gamma^3}{c^2} \vec{v} \cdot \vec{a}.$$

The Accelerated Motion

Therefore:
$$\frac{\partial \gamma}{\partial \tau} = \gamma \frac{\gamma^3}{c^2} \vec{v} \cdot \vec{a}.$$

The Derivative of $\gamma \vec{v}$ with respect to τ

$$\frac{\partial (\gamma \vec{v})}{\partial \tau} = \frac{\partial \gamma}{\partial \tau} \vec{v} + \gamma \frac{\partial \vec{v}}{\partial \tau} = \gamma \frac{\gamma^3}{c^2} \vec{v} \cdot \vec{a} \vec{v} + \gamma \frac{\partial \vec{v}}{\partial t} \frac{\partial t}{\partial \tau}$$
$$= \gamma \frac{\gamma^3}{c^2} \vec{v} \cdot \vec{a} \vec{v} + \gamma^2 \vec{a}.$$

Result

Finally it is possible to define the 4−acceleration:
$$\vec{A} = \left(\frac{\gamma^4}{c^2} \vec{v} \cdot \vec{a}, \gamma \frac{\gamma^3}{c^3} \vec{v} \cdot \vec{a} \vec{v} + \frac{1}{c} \gamma^2 \vec{a} \right). \tag{9.2}$$

9.4. The Laws of Uniformly Accelerated Motion

In first place, it is necessary to find the square norm of the 4−acceleration.

The Square Norm of 4−Acceleration

First, an hypothesis must be defined:

First hypothesis: $\vec{v} \parallel \vec{a}$, as in the case of rectilinear motion or parabolic Newtonian motion, in which in fact the vertical component of $\vec{v}$, $\vec{v}_y \parallel \vec{g}$. Therefore:
$$\vec{v}_y \cdot \vec{a} = va \cos(0 \vee \pi) = \pm va.$$

At this point:
$$\vec{A} = \left(\pm \frac{\gamma^4}{c^2} va, \frac{\gamma^4}{c^3} va(\pm 1)\vec{v} + \frac{1}{c} \gamma^2 \vec{a} \right)$$

so that the Lorentz-Minkowski square norm of $\vec{A}$ is:

$$
\begin{aligned}
|\vec{A}|^2 &= \frac{\gamma^8}{c^4}v^2a^2 - \left(\frac{\gamma^8}{c^6}v^2a^2v^2 + \frac{1}{c^2}\gamma^4a^2 + \frac{2\gamma^6}{c^4}va(\pm 1)\vec{v}\cdot\vec{a}\right) \\
&= \frac{\gamma^8}{c^4}v^2a^2 - \frac{\gamma^8}{c^6}v^2a^2v^2 - \frac{1}{c^2}\gamma^4a^2 - \frac{2\gamma^6}{c^4}v^2a^2(\pm 1)^2 \\
&= \frac{\gamma^8}{c^4}v^2a^2\left(1-\frac{v^2}{c^2}\right) - \frac{\gamma^4}{c^2}a^2\left(1+2\frac{\gamma^2}{c^2}v^2\right) \\
&= \frac{\gamma^6}{c^4}v^2a^2 - \frac{\gamma^4}{c^2}a^2\left(1+2\gamma^2\left(\frac{v}{c}\right)^2\right) = \frac{\gamma^6}{c^4}v^2a^2 - \frac{\gamma^4}{c^2}a^2 - 2\frac{\gamma^6}{c^2}a^2\frac{v^2}{c^2} \\
&= \frac{\gamma^6 v^2 a^2}{c^4} - \frac{2\gamma^6 v^2 a^2}{c^4} - \frac{\gamma^4 a^2}{c^2} = -\frac{\gamma^4 a^2}{c^2}\left(\frac{\gamma^2 v^2}{c^2}+1\right) \\
&= -\frac{\gamma^4 a^2}{c^2}\left(\frac{v^2+c^2-v^2}{c^2-v^2}\right) = -\frac{\gamma^4 a^2}{c^2}\gamma^2 = -\frac{\gamma^6}{c^2}a^2. \quad (9.3)
\end{aligned}
$$

Then:

$$|\vec{A}| = \sqrt{-\left(-\frac{\gamma^6}{c^2}a^2\right)} = \frac{\gamma^3}{c}|a|. \quad (9.4)$$

Note that it was necessary to insert the minus as real radicands must always be positive: it is a mathematical adjustment that does not affect the physical proof.

At this point the equation found can be rewritten in a different way. In order to achieve this target it is possible, under two further hypotheses, to prove the equality between $\gamma^3 a$ and $\frac{\partial(\gamma\vec{v})}{\partial t}$. First of all:

$$\frac{\partial(\gamma\vec{v})}{\partial t} = \frac{\partial\gamma}{\partial t}\vec{v} + \gamma\frac{\partial\vec{v}}{\partial t} = \frac{\gamma^3}{c^2}\left(\vec{v}\cdot\frac{\partial\vec{v}}{\partial t}\right)\vec{v} + \gamma\frac{\partial\vec{v}}{\partial t}.$$

It is now necessary to impose a further hypothesis.

Second hypothesis: $\vec{v} \parallel \vec{x} \rightarrow \vec{v} \xleftrightarrow{\cong} \pm v$. Therefore:

$$\frac{\gamma^3}{c^2}v\cdot\frac{\partial v}{\partial t}(\pm 1)v \pm \gamma\frac{\partial v}{\partial t} = \pm\gamma\frac{\partial v}{\partial t}\left(\frac{\gamma^2}{c^2}v^2+1\right) = \pm\gamma\frac{\partial v}{\partial t}\left(\frac{+v^2+c^2-v^2}{c^2-v^2}\right) = \pm\gamma\frac{\partial v}{\partial t}\left(\frac{c^2}{c^2-v^2}\right).$$

Third hypothesis. In order to continue with the proof, only the "+" sign is considered, imposing as a further hypothesis to have a uniformly *accelerated* motion with both $\vec{v}$ and the spatial acceleration $\vec{a}$ pointing in the same direction as the $\vec{x}$ axis. It follows that

$$\frac{\partial(\gamma v)}{\partial t} = +\gamma \frac{\partial v}{\partial t}\left(\frac{c^2}{c^2-v^2}\right) = \gamma \frac{\partial v}{\partial t}\gamma^2 = \gamma^3 \frac{\partial v}{\partial t},$$

as desired. Therefore:

$$|\vec{A}|_{\text{accelerated}} = \frac{\gamma^3}{c}\frac{\partial v}{\partial t} = \frac{1}{c}\frac{\partial(\gamma v)}{\partial t}. \tag{9.5}$$

If, on the contrary, a uniformly *decelerated* motion has to be obtained, it is necessary to consider the "−" in the previous equation.

Furthermore, the derivative of the speed with respect to time, as the derivative of a decreasing quantity (speed) with respect to an increasing quantity (time), will be negative:

$$A = \frac{1}{c}\frac{\partial(\gamma v)}{\partial t} < 0.$$

In conclusion, the result would be the same as that already obtained but opposite in sign:

$$|\vec{A}|_{\text{decelerated}} = \frac{\gamma^3}{c}\left|\frac{\partial v}{\partial t}\right| = -\frac{1}{c}\frac{\partial(\gamma v)}{\partial t}. \tag{9.6}$$

This shows that the equations for accelerated motion also hold for decelerated motion.

For this reason, for sake of simplicity, only the equations of uniformly accelerated motion will be demonstrated in the sequel.

The 4−Acceleration in MITCF, namely w.r.t. τ, σ

Einstein's intuitive definition of uniform acceleration is *constant acceleration in instantaneously comoving inertial frame*. Historically, however, there have been many difficulties in capturing this definition mathematically. Hereinafter some needed clarifications will be given in order to better understand Einstein's own words.

By now $\vec{A}$ was defined as a function of t and $\vec{\sigma}$. It will be useful in the following treatment to give also a precise definition of $\vec{A}(\tau,\vec{\sigma})$, namely the 4−acceleration measured *"inside"* the accelerated frame, or, more precisely, *in the MITCF*. Standing its definition - "the *momentarily* inertial tangent *comoving* frame" - at each instant τ all the spatial differentials $d\sigma^i$, which quantifies the differential deviation between the MITCF and the accelerated frame, are identically equal to zero (this translate the fact that the MITCF is *comoving* with the accelerated frame) so, since for each i the differential $d\sigma^i$ is represented by the matrix $(\partial\sigma^i/\partial\tau)$, also $\partial\sigma^i/\partial\tau$ vanish for each instant τ. In other words, since at each instant the MITCF is *comoving* with the accelerated frame (its instantaneous 4−velocity coincides with the 4−velocity of the accelerated frame) there is *not* any relative motion between the accelerated frame and the MITCF so that $\partial\sigma^i/\partial\tau$ vanishes for every $i = 1,2,3$.

Roughly speaking the 4−velocity is here seen as the limit of the deviation of the accelerated frame from the MITCF over the proper time in which such a deviation occurs. Basis vectors $\vec{e}_j$ ($j = 0,1,2,3$) belong to the accelerated frame. Here, they are seen as functions of $\Delta\tau = \tau - \tau_0$. Their corresponding coordinates measure the deviation of the accelerated frame at time τ from MITCF at τ_0. As a matter of fact, $\Delta\tau = 0$ imply that the differences $\Delta\sigma^i$ are all equal to zero. So $\Delta\tau = 0$, which is equivalent to $\tau = \tau_0$, means that the accelerated frame coincides with its MITCF at instant τ_0. So, in order to compute the velocity of both the accelerated frame and the MITCF at each instant τ_0 it suffices to compute the following derivative in 0, which means to set $\Delta\tau = 0$ which implies $\Delta\sigma(\Delta\tau) = \vec{0}$. In details, for every fixed instant τ_0, standing the fact that (i) evaluating in $\Delta\tau = 0$ is equivalent to evaluating in $\tau = \tau_0$ and (ii) since at $\tau = \tau_0$ the two frames coincide and so

$$\Delta\sigma^i(\Delta\tau) = \sigma^i(\tau) - \sigma^i(\tau_0) - 0 = \sigma^i(\tau),$$

it results:

$$\vec{V}(\Delta\tau = 0, \Delta\vec{\sigma} = \vec{0})$$
$$= \frac{\partial\vec{\Sigma}(\Delta\tau, \Delta\vec{\sigma}(\Delta\tau))}{\partial\tau}(\Delta\tau = 0) = \frac{\partial\left(\Delta\tau\vec{e}_0(\Delta\tau) + \Delta\sigma^i(\Delta\tau)\vec{e}_i(\Delta\tau)\right)}{\partial\tau}(\Delta\tau = 0)$$
$$= \frac{\partial(\tau - \tau_0)}{\partial\tau}(0)\vec{e}_0(0) + 0\frac{\partial\vec{e}_0(\Delta\tau)}{\partial\tau}(0) + \frac{\partial\left(\Delta\sigma^i(\Delta\tau)\right)}{\partial\tau}(0)\vec{e}_i(\Delta\tau) + \Delta\sigma^i(0)\frac{\partial\vec{e}_i(\Delta\tau)}{\partial\tau}(0)$$
$$= 1\vec{e}_0(\tau_0) + \frac{\partial\left(\sigma^i(\tau)\right)}{\partial\tau}(0)\vec{e}_i(\Delta\tau) + 0\frac{\partial\vec{e}_i(\Delta\tau)}{\partial\tau}(0) = \vec{e}_0(\tau_0) + 0\vec{e}_i(\Delta\tau) = \vec{e}_0(\tau_0) = (1,0,0,0).$$

The Accelerated Motion   143

So, for each instant τ, the 4-velocity of the accelerated frame, which is the same as MITCF's one is $\vec{V}(\tau) = \vec{e}_0(\tau) = (1,0,0,0)$.

The Lorentz-Minkowsky norm[6] of such a vector is obviously 1 (as it must be, see § 11.2.) or c (if, instead of homogenizing with respect to time it is chosen to homogenize with respect to spaces, see § 8.2.).

In order to compute the 4-acceleration $\vec{A}$ it is necessary to calculate the derivative of $\vec{e}_0(\tau)$ with respect to τ, namely:

$$\vec{A}(\tau,\vec{0}) = \frac{\partial \vec{e}_0}{\partial \tau}(\tau,\vec{0}).$$

Now (see § 11.1. beyond) $\frac{\partial \vec{e}_0}{\partial \tau}(\tau,\vec{0})$ is given by the following expression in which the so called *Christoffel symbols* are involved:

$$\frac{\partial \vec{e}_0}{\partial \tau}(\tau,\vec{0}) = \left(\Gamma^k_{0j} V^j e_k\right)\bigg|_{(\tau,\vec{0})} = \left(\Gamma^k_{00} V^0 e_k\right)\bigg|_{(\tau,\vec{0})} = \left(\Gamma^k_{00} e_k\right)\bigg|_{(\tau,\vec{0})}.$$

On the other hand, the Christoffel symbols are absolutely unknown at this level, so another methodology must be devised. By now note only that - in geometric units - the Christoffel symbols are dimensioned as an acceleration.

Since

$$|\vec{V}(\tau,\vec{0})|^2 = \frac{\partial \vec{V}(\tau,\vec{0})}{\partial \tau}(\tau,\vec{0}) \cdot \frac{\partial \vec{V}(\tau,\vec{0})}{\partial \tau}(\tau,\vec{0}) = 1$$

it follows that

$$\frac{\partial}{\partial \tau}|\vec{V}(\tau,\vec{0})|^2 = 2\vec{V}(\tau,\vec{0}) \cdot \frac{\partial \vec{V}(\tau,\vec{0})}{\partial \tau}(\tau,\vec{0}) = 2\vec{V}(\tau,\vec{0}) \cdot \vec{A}(\tau,\vec{0}) = 0.$$

Now, in the MITCF, $\vec{V}$ has only the first non-zero component, so this orthogonality means that in the MITCF the 4-acceleration equals to:

$$\vec{A}(\tau,\vec{0}) = \sum_{i=1}^{3} a^i e_i = (0, a^1, a^2, a^3).$$

Finally, it is now possible to add some details in order to clarify the initial Einstein's assertion according to which, in the MITCF, the 4-acceleration

---

[6]Note that - being at each instant the MITCF an *inertial* frame - the Lorentz-Minkowsky norm must apply.

is constant in each component. First of all remember that, by definition, in a uniformly accelerated motion the module of the 4−acceleration is constant, besides, the direction of its spatial components is constant itself. Let in the MITCF the four acceleration be given by $(0, a^1, a^2, a^3)$ at an instant τ_1 and $(0, b^1, b^2, b^3)$ at another instant τ_2.

It will be proved that, for each i, $a^i = b^i$. As a matter of fact, it results that the two 4−vectors must be parallel, so that there exists a constant $k \neq 0$ such that for each i, $b^i = ka^i$. On the other hand, the L-M norm of the two 4−accelerations must be the same, so

$$|\vec{B}| = k\sqrt{(a^1)^2 + (a^2)^2 + (a^3)^2} = \sqrt{(a^1)^2 + (a^2)^2 + (a^3)^2}$$

if and only if $k = 1$. This shows that in MITCF the components of the 4−acceleration are constant, as in Galilean mechanics.

Attempts to Find out the 4−Acceleration w.r.t. τ, s

Computing the 4−acceleration starting from $\vec{\Sigma} = \vec{\Sigma}(\tau, \vec{s})$ means to compute the acceleration in the frame of reference of the accelerating body (where time is shaken by τ) which measure the world around, so that, correspondingly, instead of σ the variable "s" has been employed. So, in natural units, by homogenizing as usual with respect to time:

$$\vec{\Sigma}(\tau, \vec{s}) = \tau \vec{e}_0 + \frac{1}{c}\left(s^1 \vec{e}_1 + s^2 \vec{e}_2 + s^3 \vec{e}_3\right)$$

where $\vec{e}_0, \vec{e}_1, \vec{e}_2, \vec{e}_3$ are the basis vector belonging to the accelerated frame. Hence, in geometric units, denoting s^1 with s^x and so on,

$$\begin{aligned}\vec{V}(\tau, \vec{s}) &= \frac{\partial}{\partial \tau} \vec{\Sigma}(\tau, \vec{s}) \\ &= \frac{\partial \tau}{\partial \tau} \vec{e}_0 + \tau \frac{\partial \vec{e}_0}{\partial \tau} + \frac{\partial s^x}{\partial \tau} \vec{e}_x + s^x \frac{\partial \vec{e}_x}{\partial \tau} + \frac{\partial s^y}{\partial \tau} \vec{e}_y + s^y \frac{\partial \vec{e}_y}{\partial \tau} + \frac{\partial s^z}{\partial \tau} \vec{e}_z + s^z \frac{\partial \vec{e}_z}{\partial \tau} \\ &= \vec{e}_0 + v^x \vec{e}_x + v^y \vec{e}_y + v^z \vec{e}_z + \tau \frac{\partial \vec{e}_0}{\partial \tau} + s^x \frac{\partial \vec{e}_x}{\partial \tau} + s^y \frac{\partial \vec{e}_y}{\partial \tau} + s^z \frac{\partial \vec{e}_z}{\partial \tau}.\end{aligned}$$

On the other hand, what about the derivative with respect to τ of $\vec{e}_0, \vec{e}_x, \vec{e}_y$ and $\vec{e}_z$? It is only known that $|\vec{V}|$ must be equal to 1 and, for every $i = 0, 1, 2, 3$ (see § 11.1.):

$$\frac{\partial \vec{e}_i}{\partial \tau} = \Gamma_{ij}^k V^j \vec{e}_k = \Gamma_{ij}^k \frac{\partial \Sigma^j}{\partial \tau} \vec{e}_k = \Gamma_{i0}^k \vec{e}_k + \sum_{j=1}^{3} \Gamma_{ij}^k v^j \vec{e}_k.$$

The Accelerated Motion

So - in this particular frame - the 4−velocity depends on both v^x, v^y, v^z, namely $\vec{v}$, as in the frame with coordinates (t,σ), see (8.2), and on the derivatives of $\vec{e}_i$ with respect to the proper time, which, in turn, deeply depend on the metric, since the Christoffel symbols are a particular function of it and its inverse.

Now, it doesn't seem very relevant to go on in computing the further derivative with the aim to calculate the 4−acceleration since it will depend itself on the Christoffel symbols and so on the metric.

On the other hand, note that if $(\vec{e}_0, \vec{e}_1, \vec{e}_2, \vec{e}_3)$ denotes a basis for the non-accelerating frame it results that $\frac{\partial \vec{e}_i}{\partial \tau} = \gamma \frac{\partial \vec{e}_i}{\partial t} = \vec{0}$ for every $i = 0, \ldots, 3$. This is the reason why the computations performed in § 9.3. works properly by deriving each component singularly, without taking care at all to derive the basis vectors themselves.

In conclusion: whatever the frame of reference is, the module of $\vec{A}$ in that frame must be the same, this in virtue of the tensorial character of $\vec{A}$. That is

$$\left|\vec{A}(t,\vec{\sigma})\right| = \gamma^3 \left|\vec{a}(t)\right| = \text{constant}$$

implies, in the notations above introduced

$$\left|\vec{A}(\tau,\vec{\sigma})\right| = \left|\vec{A}(\tau,\vec{s})\right| = \text{the same constant.}$$

Law of Motion and Its Worldline

First of all, it is useful to report the three hypotheses under which it is possible to prove the equations. Besides, an addition hypothesis is added in order to measure accelerations in natural units.

- **First hypothesis**: $\vec{v} \parallel \vec{a}$

- **Second hypothesis**: $\vec{v} \parallel \vec{x} \to \vec{v} \Leftrightarrow \pm v$

- **Third hypothesis**: Uniformly *accelerated* motion.

- **Fourth hypothesis**: Natural (that is, non-geometrical) units.

In particular, under the last assumptions the modulus with sign of $\vec{A}$ that is A, equals to $\frac{\partial(\gamma v)}{\partial t}$ instead of $\frac{1}{c}\frac{\partial(\gamma v)}{\partial t}$. So

$$A = \frac{\partial(\gamma v)}{\partial t} \Rightarrow \partial(\gamma v) = A\partial t \Rightarrow \int \partial(\gamma v) = \int A\,\partial t \iff \gamma v = At + k$$

for some constant $k \in \mathbb{R}$. On the other hand

$$(\gamma v)_{|t=0} = \gamma(v_0) \cdot v(0) = k \Longrightarrow \gamma v = At + v_0\gamma(v_0).$$

So

$$\frac{v(t)}{\sqrt{1 - \left(\frac{v(t)}{c}\right)^2}} = At + v_0\gamma(v_0)$$

if and only if

$$v(t) = At\sqrt{1 - \left(\frac{v(t)}{c}\right)^2} + v_0\frac{\sqrt{1-\left(\frac{v(t)}{c}\right)^2}}{\sqrt{1-\left(\frac{v_0}{c}\right)^2}} = \frac{At}{\gamma(v(t))} + v_0\frac{\gamma(v_0)}{\gamma(v(t))}$$

therefore:

$$v(t) = \frac{1}{\gamma(v(t))}\left(At + v_0\gamma(v_0)\right). \tag{9.7}$$

It follows that

$$v^2(t) = \left(1 - \frac{v^2(t)}{c^2}\right)(At + v_0\gamma(v_0))^2$$

which, for sake of simplicity, by putting $B(t) \stackrel{\text{def}}{=} At + v_0\gamma(v_0)$ and so

$$B^2 = (At + v_0\gamma(v_0))^2$$

yields

$$v^2 = B^2 - \frac{B^2}{c^2}v^2 \iff v^2\left(1 + \frac{B^2}{c^2}\right) = B^2 \iff v^2 = \frac{B^2}{1 + \left(\frac{B}{c}\right)^2}$$

that is
$$v(t) = \frac{At + v_0\gamma(v_0)}{\sqrt{1 + \left(\frac{At+v_0\gamma(v_0)}{c}\right)^2}} \qquad (9.8)$$

First of all note that for $t = 0$

$$v(0) = v_0\gamma(v_0) \cdot \frac{1}{\sqrt{1 + \left(\frac{v_0\gamma(v_0)}{c}\right)^2}} = v_0\gamma(v_0) \cdot \frac{1}{\gamma(v_0)} = v_0$$

as desired, besides $\lim_{t \to +\infty} v(t) = c$, which is consistent with the assumption that σ is the proper distance measured in an *inertial* frame. It follows that the derivative of v, namely the scalar acceleration a, must tend to zero as $t \to +\infty$, as shown here and, in a faster, even if less constructively way, in § 9.6.:

$$a(t) = \frac{\partial v}{\partial t} = \frac{A}{\left(1 + \left(\frac{At+v_0\gamma(v_0)}{c}\right)^2\right)^{\frac{3}{2}}} \qquad (9.9)$$

so that $\lim_{t \to +\infty} a(t) = 0$, which is consistent with the postulates.

Now it is possible to deduce the *law of motion*: remembering that $v(t) = \partial\sigma/\partial t$ it results:

$$\int \partial\sigma = \int (At + v_0\gamma(v_0))\left(1 + \left(\frac{At+v_0\gamma(v_0)}{c}\right)^2\right)^{-\frac{1}{2}} \partial t$$

if and only if

$$\sigma(t) = \int \left(\frac{A^2t^2}{c^2} + \frac{2Av_0\gamma(v_0)t}{c^2} + \frac{v_0^2\gamma^2(v_0)}{c^2} + 1\right)^{-\frac{1}{2}} (At + v_0\gamma(v_0))\, \partial t$$

$$= \frac{c^2}{2A}\int \left(\frac{A^2t^2}{c^2} + \frac{2Av_0\gamma(v_0)t}{c^2} + \frac{v_0^2\gamma^2(v_0)}{c^2} + 1\right)^{-\frac{1}{2}} \cdot \frac{2A}{c^2}(At + v_0\gamma(v_0))\, \partial t$$

$$= \frac{c^2}{2A}\left(\frac{A^2t^2}{c^2} + \frac{2Av_0\gamma(v_0)t}{c^2} + \frac{v_0^2\gamma^2(v_0)}{c^2} + 1\right)^{\frac{1}{2}} \cdot 2 + k$$

$$= \frac{c^2}{A}\left(\frac{A^2t^2}{c^2} + \frac{2Av_0\gamma(v_0)t}{c^2} + \frac{v_0^2\gamma^2(v_0)}{c^2} + 1\right)^{\frac{1}{2}} + k.$$

In order to determine the value of the constant k, without loss of generality let $\sigma(0) = 0$ so that

$$\sigma(0) = \frac{c^2}{A}\sqrt{\frac{v_0^2\gamma^2(v_0)}{c^2}+1}+k = 0 \iff k = -\frac{c^2}{A}\sqrt{\frac{v_0^2\gamma^2(v_0)}{c^2}+1} = -\frac{c^2}{A}\gamma(v_0)$$

therefore:

$$\sigma(t) = \frac{c^2}{A}\sqrt{\frac{A^2t^2}{c^2}+\frac{2Av_0\gamma(v_0)t}{c^2}+\frac{v_0^2\gamma^2(v_0)}{c^2}+1}-\frac{c^2}{A}\gamma(v_0).$$

The **law of motion** is then:

$$\boxed{\sigma(t) = \frac{c^2}{A}\sqrt{1+\left(\frac{At+v_0\gamma(v_0)}{c}\right)^2}-\frac{c^2}{A}\gamma(v_0)} \qquad (9.10)$$

The equation above describes the *proper position* measured from an inertial frame expressed as a function of the *non proper* time t.

Graph of the law of motion: worldline. In the $\left(\frac{\sigma}{c},t\right)$ plane, the graph of the mapping $\frac{\sigma}{c}: t \mapsto \frac{\sigma(t)}{c}$ with σ as in (9.10) is the equilateral hyperbola's right branch

$$\frac{(\sigma/c)^2}{\left(\frac{c}{A}\right)^2}-\frac{t^2}{\left(\frac{c}{A}\right)^2}=1,\ \sigma \geq 0 \iff \frac{\sigma}{c}(t) = \frac{c}{A}\sqrt{1+\left(\frac{At}{c}\right)^2} \qquad (9.11)$$

translated by the vector

$$\left(-\frac{c\gamma(v_0)}{A},-\frac{v_0\gamma(v_0)}{A}\right). \qquad (9.12)$$

As a matter of fact the asymptotes of the equilateral hyperbola

$$\frac{(\sigma/c)^2}{\left(\frac{c}{A}\right)^2}-\frac{t^2}{\left(\frac{c}{A}\right)^2}=1$$

are just the bisectors $t = \pm\frac{\sigma}{c}$. They constitute the light cone with vertex coincident with the origin.

The Accelerated Motion

Note that the vertex of the branch of equilateral hyperbola given by the equation 9.10 is the vertex of (9.11), i.e. $\left(\frac{c}{A}, 0\right)$, translated by the vector 9.12, namely:

$$\mathcal{V} = \left(\frac{\sigma}{c}(\mathcal{V}), t(\mathcal{V})\right) = \left(\frac{c}{A}(1 - \gamma(v_0)), -\frac{v_0}{A}\gamma(v_0)\right).$$

What about the speed for $t = t(\mathcal{V})$? Let's substitute in (9.8):

$$v\left(-\frac{v_0}{A}\gamma(v_0)\right) = \frac{-v_0\gamma(v_0) + v_0\gamma(v_0)}{\sqrt{1 + \left(\frac{-v_0\gamma(v_0) + v_0\gamma(v_0)}{c}\right)^2}} = 0. \tag{9.13}$$

On the other hand, this was surely a *foreseeable result*. To show this, summarising, given the function:

$$t \longmapsto \frac{\sigma}{c}(t) = \frac{c}{A}\sqrt{1 + \left(\frac{At + v_0\gamma(v_0)}{c}\right)^2} - \frac{c}{A}\gamma(v_0)$$

its graph in the $\left(\frac{\sigma}{c}, t\right)$ plane consists in the right branch of equilateral hyperbola whose asymptotes are the lines $t + \frac{v_0\gamma(v_0)}{A} = \pm\left(\frac{\sigma}{c} + \frac{c\gamma(v_0)}{A}\right)$, namely

$$t = \pm\frac{\sigma}{c} + \frac{\gamma(v_0)}{A}(\pm c - v_0)$$

so that, in conclusion:

- if $t < -\frac{v_0\gamma(v_0)}{A}$ then we are on the lower-right semi-branch of the given equilateral hyperbola: t increases implies $\frac{\sigma}{c}$ decreases, that is we are *approaching* to the vertex $\mathcal{V}$;

- if $t > -\frac{v_0\gamma(v_0)}{A}$ then we are *walking* on the the upper-right semi-branch of the equilateral hyperbola: t increases implies $\frac{\sigma}{c}$ increases, namely we are *departing* from the vertex $\mathcal{V}$;

- standing the two points above, if $t = -\frac{v_0\gamma(v_0)}{A}$ then the motion reverses its direction, we are *instantaneously* in $\mathcal{V}$ and, for the continuity of the function speed, $v\left(\frac{v_0\gamma(v_0)}{A}\right)$ must be equal to zero, as already computed in (9.13);

- standing the three points above it follows that the motion is always accelerated towards the positive σ/c axis. In particular, if $t < -v_0\gamma(v_0)/A$ then $v < 0$ so that the velocity is anti-parallel with respect to the $\frac{\sigma}{c}$ axis: the motion is *decelerated*. On the contrary, if $t > -v_0\gamma(v_0)/A$ then $v > 0$ and the motion is *accelerated*. Note that in every circumstances (both for $v < 0$ and $v > 0$) it results that $\partial v/\partial t > 0$, as a matter of fact, $v < 0$ implies $v \nearrow 0$, on the other hand $v > 0$ implies $v \nearrow c$ which means that whatever the sign of v is, it results that $\partial v/\partial t > 0$;

- $v_0 < 0$ means that for $t = 0$ the accelerating body is approaching to the hyperbola's vertex $\mathcal{V}$; if $v_0 > 0$ means that for $t = 0$ the body is departing from the vertex, $v_0 = 0$ means that for $t = 0$ the body is in the hyperbola's vertex.

Standing the above it follows that in every circumstances, whatever the sign of v_0 (including $v_0 = 0$) it results that:

$$v(t) = \frac{At + v_0\gamma(v_0)}{\sqrt{1 + \left(\frac{At+v_0\gamma(v_0)}{c}\right)^2}}. \qquad (9.14)$$

Note that in the discussion above a sign as been given both to A and v_0 (A always positive, v_0 depending on the starting point). Summarizing:

$$v_0 < 0 \Rightarrow \begin{cases} t < \left|\frac{v_0\gamma(v_0)}{A}\right| & \Rightarrow \left(t \nearrow \Rightarrow \frac{\sigma}{c} \searrow \text{ i.e. } approaching \text{ to } \mathcal{V}\right), \\ t = \left|\frac{v_0\gamma(v_0)}{A}\right| & \Rightarrow \text{ being in } \mathcal{V}, \\ t > \left|\frac{v_0\gamma(v_0)}{A}\right| & \Rightarrow \left(t \nearrow \Rightarrow \frac{\sigma}{c} \nearrow \text{ i.e. } departing \text{ from } \mathcal{V}\right); \end{cases}$$

$v_0 = 0 \Rightarrow t = 0 \Rightarrow$ being *instantaneously* in $\mathcal{V} = (0,0)$;

$$v_0 > 0 \Rightarrow t > -\left|\frac{v_0\gamma(v_0)}{A}\right| \Rightarrow \left(t \nearrow \Rightarrow \frac{\sigma}{c} \nearrow \text{ i.e. } departing \text{ from } \mathcal{V}\right).$$

The Accelerated Motion   151

Analysis in terms of proper time. On the one hand, equation (9.10) can be easily transformed into

$$\left(t+\frac{v_0\gamma(v_0)}{A}\right)^2 - \left(\frac{\sigma}{c}+\frac{c}{A}\gamma(v_0)\right)^2 = -\frac{c^2}{A^2}. \quad (9.15)$$

On the other hand, as per equations 9.3 and 9.4, in natural units the Lorentz-Minkowsky's squared norm of the 4-acceleration $\vec{A} = \left(\frac{\partial^2 t}{\partial \tau^2}, \frac{\partial^2 \sigma}{\partial \tau^2}\right)$ is[7]:

$$\left(\frac{\partial^2 t}{\partial \tau^2}\right)^2 - \left(\frac{\partial^2 \sigma}{c\partial \tau^2}\right)^2 = -\frac{A^2}{c^2}. \quad (9.16)$$

Putting the equations 9.15 and 9.16 together it follows that

$$\left(\frac{\partial^2 t}{\partial \tau^2}\right)^2 - \left(\frac{\partial^2 \sigma}{c\partial \tau^2}\right)^2 = \frac{A^4}{c^4}\left(\left(t+\frac{v_0\gamma(v_0)}{A}\right)^2 - \left(\frac{\sigma}{c}+\frac{c\gamma(v_0)}{A}\right)^2\right)$$

if and only if

$$\left(\frac{\partial^2 t}{\partial \tau^2}\right)^2 - \left(\frac{\partial^2 \sigma}{c\partial \tau^2}\right)^2 = \left(\frac{A^2}{c^2}\left(t+\frac{v_0\gamma(v_0)}{A}\right)\right)^2 - \left(\frac{A^2}{c^2}\left(\frac{\sigma}{c}+\frac{c\gamma(v_0)}{A}\right)\right)^2$$

or equivalently

$$\left(\frac{\partial^2 t}{\partial \tau^2}\right)^2 - \left(\frac{A^2}{c^2}\left(t+\frac{v_0\gamma(v_0)}{A}\right)\right)^2 = \left(\frac{\partial^2 \sigma}{c\partial \tau^2}\right)^2 - \left(\frac{A^2}{c^2}\left(\frac{\sigma}{c}+\frac{c\gamma(v_0)}{A}\right)\right)^2$$

which, for the linear independence of the time-coordinate from the spatial one, both the first and the second member of the equation above must be equal to zero, so that it is possibile to write the following system of two decoupled non-homogeneous differential equations of the second order:

$$\begin{cases} \frac{\partial^2 t}{\partial \tau^2} = \frac{A^2}{c^2}\left(t+\frac{v_0\gamma(v_0)}{A}\right) \\ \frac{\partial^2 \sigma}{c\partial \tau^2} = \frac{A^2}{c^2}\left(\frac{\sigma}{c}+\frac{c\gamma(v_0)}{A}\right) \iff \frac{\partial^2 \sigma}{\partial \tau^2} = \frac{A^2}{c^2}\left(\sigma+\frac{c^2\gamma(v_0)}{A}\right) \end{cases} \quad (9.17)$$

---

[7] Here and after both the y and the z coordinates will be omitted, which are assumed to be constant, as per the assumptions of § 9.4.

with initial conditions *(i)* $t(0) = 0$, *(ii)* $\frac{\partial t}{\partial \tau}(0) = \gamma|_{\tau=0=t} = \gamma(v_0)$; *(iii)* $\sigma(0) = 0$ and

(iv) $\frac{\partial \sigma}{\partial \tau}(0) = \frac{\partial \sigma}{\partial t}\bigg|_{t(\tau_0)} \frac{\partial t}{\partial \tau}\bigg|_{\tau=0} = \frac{\partial \sigma}{\partial t}\bigg|_{t=0} \frac{\partial t}{\partial \tau}\bigg|_{\tau=0} = v|_{t=0}\gamma|_{t=0} = v_0\gamma(v_0).$

By the *Cauchy–Lipschitz* theorem such a differential system with assigned initial conditions has a unique solution, which can be found, for example, by the method of undetermined coefficients:

$$\begin{cases} t(\tau) = \frac{\gamma(v_0)}{A}\left(v_0 \cosh\left(\frac{A}{c}\tau\right) + c \sinh\left(\frac{A}{c}\tau\right) - v_0\right) \\ \sigma(\tau) = \frac{c\gamma(v_0)}{A}\left(c \cosh\left(\frac{A}{c}\tau\right) + v_0 \sinh\left(\frac{A}{c}\tau\right) - c\right). \end{cases} \quad (9.18)$$

Note that both σ and τ are proper: σ is measured from the inertial frame, τ is measured in the non-inertial one, for example, an accelerated spaceship.

Deriving with respect to the proper time it follows:

$$\begin{cases} \frac{\partial t}{\partial \tau}(\tau) = \gamma(v(\tau)) = \gamma(v_0)\left(\frac{v_0}{c}\sinh\left(\frac{A}{c}\tau\right) + \cosh\left(\frac{A}{c}\tau\right)\right) \\ \frac{\partial \sigma}{\partial \tau}(\tau) = \frac{\partial \sigma}{\partial t}(t)\frac{\partial t}{\partial \tau}(\tau) = v(t)\gamma(v(\tau)) = \gamma(v_0)\left(c \sinh\left(\frac{A}{c}\tau\right) + v_0 \cosh\left(\frac{A}{c}\tau\right)\right) \end{cases} \quad (9.19)$$

besides the second derivatives are:

$$\begin{cases} \frac{\partial^2 t}{\partial \tau^2}(\tau) = \frac{A\gamma(v_0)}{c}\left(\frac{v_0}{c}\cosh\left(\frac{A}{c}\tau\right) + \sinh\left(\frac{A}{c}\tau\right)\right) \\ \frac{\partial^2 \sigma}{\partial \tau^2}(\tau) = \frac{A\gamma(v_0)}{c}\left(c \cosh\left(\frac{A}{c}\tau\right) + v_0 \sinh\left(\frac{A}{c}\tau\right)\right). \end{cases} \quad (9.20)$$

During the uniform acceleration, at different instants t and τ which are related by the first of (9.18), surely the relative velocity coincides:

$$\begin{cases} v(\tau) = v(t) = \frac{At + v_0\gamma(v_0)}{\sqrt{1 + \left(\frac{At + v_0\gamma(v_0)}{c}\right)^2}} \\ t(\tau) = \frac{\gamma(v_0)}{A}\left(v_0 \cosh\left(\frac{A}{c}\tau\right) + c \sinh\left(\frac{A}{c}\tau\right) - v_0\right). \end{cases}$$

So, substituting, the following result is achieved:

$$v(\tau) = \frac{\gamma(v_0)\left[v_0 \cosh\left(\frac{A}{c}\tau\right) + c \sinh\left(\frac{A}{c}\tau\right)\right]}{\sqrt{1 + \left(\frac{\gamma(v_0)\left[v_0 \cosh\left(\frac{A}{c}\tau\right) + c \sinh\left(\frac{A}{c}\tau\right)\right]}{c}\right)^2}} \quad (9.21)$$

The Accelerated Motion

Deriving with respect to τ, the spatial acceleration measured by someone or something which is uniformly accelerating is then:

$$a(\tau) = \frac{\partial v}{\partial \tau}(\tau) = \frac{\gamma(v_0)\frac{A}{c}\left[v_0 \sinh\left(\frac{A}{c}\tau\right) + c\cosh\left(\frac{A}{c}\tau\right)\right]}{\left[1 + \left(\frac{\gamma(v_0)\left[v_0\cosh\left(\frac{A}{c}\tau\right)+c\sinh\left(\frac{A}{c}\tau\right)\right]}{c}\right)^2\right]^{\frac{3}{2}}} = \frac{\frac{\partial^2 \sigma}{\partial \tau^2}(\tau)}{\left(1 + \left(\frac{\partial \sigma}{c\partial \tau}\right)^2\right)^{\frac{3}{2}}}.$$

(9.22)

In particular, $\lim_{\tau \to +\infty} a(\tau) = 0$ so that $\lim_{\tau \to +\infty} v(\tau) = c$, which is consistent with the second postulate of the theory.

Exercise 18. *Show that* $\lim_{\tau \to +\infty} v(\tau) = c$ *and* $\lim_{\tau \to +\infty} a(\tau) = 0$.

Solution: It suffices to remember that $1 + x^2 = O(x^2)$. $\square$

An Interesting Link with the SEP

Thanks to SEP an astronaut in a spaceship with his feet comfortably in contact with its floor could think to leave steady in a planet whose gravity at time τ equals $|\vec{A}(\tau,s)|$, which is equal to[8] $|\vec{A}(t,\sigma)| = |\vec{A}(\tau,\sigma)| = |(0,a^1,a^2,a^3)| \equiv A$. At each proper time τ, also an accelerometer could not establish if the astronaut is moving with a relativistic acceleration equal to A or if he is leaving on a planet whose gravity at each time τ is $g = A$.

Law of motion in terms of proper time In general, for every v_0, despite the existence of the primitive of $v = v(\tau)$, it's very hard both to compute and to write it, so it is preferred to give an explicit expression of the law of motion only for $v_0 = 0$. As a matter of fact, $v_0 = 0$ imply:

$$\frac{\partial s}{\partial \tau}(\tau) = v(\tau) = \frac{c\sinh\left(\frac{A}{c}\tau\right)}{\sqrt{1+\sinh^2\left(\frac{A}{c}\tau\right)}} = \frac{c\sinh\left(\frac{A}{c}\tau\right)}{\cosh\left(\frac{A}{c}\tau\right)} = c\tanh\left(\frac{A}{c}\tau\right)$$

so

$$\int \partial s = \frac{c^2}{A}\int \frac{\frac{A}{c}\sinh\left(\frac{A}{c}\tau\right)}{\cosh\left(\frac{A}{c}\tau\right)}\partial \tau = \frac{c^2}{A}\ln\left(\cosh\left(\frac{A}{c}\tau\right)\right) + k. \quad (9.23)$$

---

[8] Regarding this last equation remember § 9.4.

Now, if, as usual $s(0) = 0$ so that $k = 0$, under the hypothesis $v_0 = 0$, the *law of motion* in terms of proper time is:

$$\boxed{s(\tau) = \frac{c^2}{A} \ln\left(\cosh\left(\frac{A}{c}\tau\right)\right)} \qquad (9.24)$$

If $a(\tau)$ is constant In order to prepare the groundwork for a deep analysis of the so famous *twin paradox* treated in the right context of the accelerated motion the analytic expression of $\tau = \tau(t)$ will now be deduced. Instead of inverting the first equation of (9.18) which would require some tedious computations, starting from the differential time dilation relation and taking into account the equation 9.8 for the right expression of v, it follows immediately that

$$\partial t = \frac{\partial \tau}{\sqrt{1-\left(\frac{v}{c}\right)^2}} \Rightarrow \partial \tau = \sqrt{1-\left(\frac{At+v_0\gamma(v_0)}{c\sqrt{1+\left(\frac{At+v_0\gamma(v_0)}{c}\right)^2}}\right)^2}\, \partial t = \sqrt{1-\frac{\left(\frac{At+v_0\gamma(v_0)}{c}\right)^2}{1+\left(\frac{At+v_0\gamma(v_0)}{c}\right)^2}}\, \partial t$$

therefore:

$$\int \partial \tau = \int \left(1+\left(\frac{At+v_0\gamma(v_0)}{c}\right)^2\right)^{-\frac{1}{2}} \partial t.$$

Now let

$$r := \frac{At+v_0\gamma(v_0)}{c} \rightsquigarrow \partial r = \frac{A}{c}\partial t$$

and so

$$\tau = \frac{c}{A}\int \frac{\partial r}{\sqrt{1+r^2}}.$$

Define z such that $r = \sinh z$ so that $\partial r = \cosh z\, \partial z$ and finally, standing the fact that $\cosh^2 z - \sinh^2 z = 1$,

$$\sqrt{1+r^2} = \sqrt{1+\sinh^2 z} = |\cosh(z)| = \cosh z.$$

So

$$\tau = \frac{c}{A}\int \frac{\cosh z}{\cosh z}\, dz = \frac{c}{A}\int dz = \frac{c}{A}z + k = \frac{c}{A}\operatorname{arsinh}(r) + k = \frac{c}{A}\operatorname{arsinh}\left(\frac{At+v_0\gamma(v_0)}{c}\right) + k.$$

In particular:

$$\tau(0) = 0 \Longrightarrow \frac{c}{A}\operatorname{arsinh}\left(\frac{v_0\gamma(v_0)}{c}\right) + k = 0 \iff k = -\frac{c}{A}\operatorname{arsinh}\left(\frac{v_0\gamma(v_0)}{c}\right).$$

The Accelerated Motion

Under the assumption that $\tau_{|t=0} = 0$ the equation of uniformly accelerated relativistic motion for proper time is presented below:

$$\boxed{\tau(t) = \frac{c}{A}\operatorname{arsinh}\left(\frac{At + v_0\gamma(v_0)}{c}\right) - \frac{c}{A}\operatorname{arsinh}\left(\frac{v_0\gamma(v_0)}{c}\right)} \qquad (9.25)$$

Crosscheck. From the previous (9.25) it is easy to deduce

$$t(\tau) = \frac{c}{A}\sinh\left(\frac{A}{c}\tau + \operatorname{arsinh}\left(\frac{v_0\gamma(v_0)}{c}\right)\right) - \frac{v_0\gamma(v_0)}{A} \qquad (9.26)$$

which need to be equal to the first relation of (9.18). This is true and reduces to a straight check by using the following results: (i) for every $x \in \mathbb{R}$ it results that $\cosh(\operatorname{arsinh}(x)) = \sqrt{x^2 + 1}$, (ii) the addition formula for sinh.

Relationships between Proper and Non-Proper Spaces

The relationship between proper and non-proper spaces will be now deduced assuming that the initial speed is equal to zero: $v_0 = v(0) = 0$. Starting from (9.24) it follows that

$$\tau = \frac{c}{A}\operatorname{arcosh}\left(e^{\frac{A}{c^2}s}\right)$$

so, substituting in (9.18) with $v_0 = 0$:

$$\sigma(s) = \frac{c}{A}\left(c\cosh\left(\frac{A}{c}\frac{c}{A}\operatorname{arcosh}\left(e^{\frac{A}{c^2}s}\right)\right) - c\right) = \frac{c^2}{A}\left(e^{\frac{A}{c^2}s} - 1\right)$$

hence

$$\boxed{\sigma(s) = \frac{c^2}{A}\left(e^{\frac{A}{c^2}s} - 1\right)} \qquad (9.27)$$

9.5. Equation Summary

Uniformly Accelerated & Decelerated Motion in Natural Units

- $\left|\vec{A}\right| = |A| = \left|\frac{\partial(\gamma v)}{\partial t}\right| = \gamma^3 |a| = \gamma^3 \left|\frac{\partial v}{\partial t}\right| = \left|\frac{\partial(\gamma v)}{\gamma \partial \tau}\right| \equiv \text{constant}$

- $t(\tau) = \frac{v_0 \gamma(v_0)}{A} \cosh\left(\frac{A}{c}\tau\right) + \frac{c\gamma(v_0)}{A} \sinh\left(\frac{A}{c}\tau\right) - \frac{v_0 \gamma(v_0)}{A}$

- $t(\tau) = \frac{c}{A} \sinh\left(\frac{A}{c}\tau + \operatorname{arsinh}\left(\frac{v_0 \gamma(v_0)}{c}\right)\right) - \frac{v_0 \gamma(v_0)}{A}$

- $\tau(t) = \frac{c}{A} \operatorname{arsinh}\left(\frac{At + v_0 \gamma(v_0)}{c}\right) - \frac{c}{A} \operatorname{arsinh}\left(\frac{v_0 \gamma(v_0)}{c}\right)$

- $\sigma(t) = \frac{c^2}{A} \sqrt{1 + \left(\frac{At + v_0 \gamma(v_0)}{c}\right)^2} - \frac{c^2}{A} \gamma(v_0)$

- $s(\tau) = \frac{c^2}{A} \ln\left(\cosh\left(\frac{A}{c}\tau\right)\right)$ (under the hypothesis $v_0 = 0$)

- $\sigma(s) = \frac{c^2}{A} \left(e^{\frac{A}{c^2}s} - 1\right)$ (under the hypothesis $v_0 = 0$)

- $v(t) = \frac{1}{\gamma(v)} (At + v_0 \gamma(v_0)) = \frac{At + v_0 \gamma(v_0)}{\sqrt{1 + \left(\frac{At + v_0 \gamma(v_0)}{c}\right)^2}}$

- $v(\tau) = \frac{\gamma(v_0)\left(v_0 \cosh\left(\frac{A}{c}\tau\right) + c \sinh\left(\frac{A}{c}\tau\right)\right)}{\sqrt{1 + \left(\frac{\gamma(v_0)\left(v_0 \cosh\left(\frac{A}{c}\tau\right) + c \sinh\left(\frac{A}{c}\tau\right)\right)}{c}\right)^2}}$

- $a(t) = \dfrac{A}{\left(1 + \left(\frac{At + v_0 \gamma(v_0)}{c}\right)^2\right)^{\frac{3}{2}}}$

- $a(\tau) = \dfrac{\gamma(v_0)\frac{A}{c}\left[v_0 \sinh\left(\frac{A}{c}\tau\right) + c \cosh\left(\frac{A}{c}\tau\right)\right]}{\left[1 + \left(\frac{\gamma(v_0)\left[v_0 \cosh\left(\frac{A}{c}\tau\right) + c \sinh\left(\frac{A}{c}\tau\right)\right]}{c}\right)^2\right]^{3/2}} = \dfrac{\frac{\partial^2 \sigma}{\partial \tau^2}(\tau)}{\left(1 + \left(\frac{\partial \sigma}{c \partial \tau}\right)^2\right)^{3/2}}$

All the equations above apply both for *accelerated* and *decelerated* motion. In any circumstances, after having fixed a one-dimensional system of coordinates, the signs of both A and v_0 have to be settled accordingly.

9.6. Boundary Analysis

This section's aim is to analyse what happens when speed tends to 0 or ∞. It's fundamental to remind that the same hypothesis defined in 9.4. apply. In particular, it is necessary to consider natural (non-geometrical) units.

When Speed Is Much Smaller than c

In natural units, as $A = \frac{\partial(\gamma v)}{\partial t} = \gamma^3 \frac{\partial v}{\partial t} = \gamma^3 a$, $(v \ll c \vee v_0 \ll c) \Rightarrow \gamma \to 1 \Rightarrow A \to a$. Therefore it's possible to derive the non-relativistic equations of uniformly accelerated motion:

- $v(t) = \frac{1}{\gamma(v)}(At + v_0\gamma(v_0)) \approx \frac{1}{1}(at + v_0) = at + v_0$;

- besides, standing the fact that both $\frac{at+v_0}{c}, \frac{v_0}{c} \ll 1$, by using Taylor expansions centered at zero $(1+x)^{\frac{1}{2}} \approx 1 + \frac{1}{2}x$ it follows that:

$$\sigma(t) = \frac{c^2}{A}\left(\sqrt{1 + \left(\frac{At + v_0\gamma(v_0)}{c}\right)^2} - \sqrt{1 + \left(\frac{v_0\gamma(v_0)}{c}\right)^2}\right)$$

$$\approx \frac{c^2}{a}\left(\sqrt{1 + \left(\frac{at + v_0}{c}\right)^2} - \sqrt{1 + \left(\frac{v_0}{c}\right)^2}\right)$$

$$\approx \frac{c^2}{a}\left(1 + \frac{1}{2}\left(\frac{at+v_0}{c}\right)^2 - 1 - \frac{1}{2}\left(\frac{v_0}{c}\right)^2\right) = \frac{c^2}{a}\left(\frac{a^2t^2 + 2atv_0}{2c^2}\right) = \frac{1}{2}at^2 + v_0t$$

which coincides with the Newtonian's limit, and the corresponding principle is safe.

When Speed Tends to c

Standing the fact that

$$A = \frac{\gamma^3}{c}a$$

if $v \to c$ and therefore $\gamma \to +\infty$ then, assuming uniformly accelerated motion, a must tend to zero. In other words, the bigger the velocity is (always limited by c) the smaller the ordinary spatial acceleration a become, that is the uniformly accelerated motion tends to uniform motion.

9.7. Still Parabolic Motion? No, Hyperbolic!

First of all let's rewrite the hypothesis of section 9.4. in such a manner that the y axis is, as usual, vertically oriented upward and the x axis is horizontally oriented to the right, with the scalar acceleration $\vec{a}$ parallel to y. Now, let $v_{oy} := v_o$ and suppose the initial velocity to have also a component v_{ox} directed along the x axis, so that:

$$\begin{cases} \sigma_x(t) = v_{ox} t \\ \sigma_y(t) = \frac{c^2}{A}\left(\sqrt{1+\left(\frac{At+v_{oy}\gamma(v_{oy})}{c}\right)^2} - \gamma(v_{oy})\right). \end{cases} \quad (9.28)$$

The above equations are parametrized by the non-proper time t.

As in Newtonian's motion, standing the fact that uniform spatial acceleration $\vec{a}$ is directed along the y axis, one could suppose that equations 9.28 describe a parabolic trajectory whose axis is parallel to y. As a matter of fact, they represent an equilateral hyperbola. Indeed, by substituting $t = \frac{\sigma_x}{v_{ox}}$ in the second equation (note that v_{ox} must be different from 0, if and only if $\theta \neq \pm\pi/2$, where θ denotes the angle of elevation or depression), the following Cartesian's equation follows immediately:

$$\sigma_y(\sigma_x) = \frac{c^2}{A}\left(\sqrt{1+\left(\frac{A\sigma_x}{cv_{ox}}+\frac{v_{oy}\gamma(v_{oy})}{c}\right)^2} - \gamma(v_{oy})\right) \quad (9.29)$$

which represents in the $\sigma_x\sigma_y$ plane the hyperbola's upper branch

$$\sigma_y(\sigma_x) = \frac{c^2}{A}\sqrt{1+\left(\frac{A\sigma_x}{cv_{ox}}\right)^2} \quad (9.30)$$

translated by the vector

$$\begin{pmatrix} -\frac{v_{oy}\gamma(v_{oy})}{c} \\ -\frac{c^2}{A}\gamma(v_{oy}) \end{pmatrix}.$$

Note that the above mentioned hyperbola is equilateral if and only if v_{0x} equals c (and so $v_{oy} = 0$).

Problem 2. *Consider the following gedankenexperiment[9]: there is a falling elevator, that is an elevator cabin that is in free fall, and assume*

---

[9]thought experiment, term coined by the Danish physicist and chemist Hans Christian Ørsted.

that, at the start of our experiment, the elevator is still at rest. At this exact instant, someone or something shoots a brief pulse of light into the elevator through a little hole in the elevator's side. Our pulse of light is aimed in such a way that the light travels horizontally as it enters the elevator. Besides the elevator starts its free fall at the precise instant that the light pulse enters it. (a) How is the light's trajectory viewed from an outside reference frame mathematically characterize? (b) Suppose there is a second hole in the elevator wall, straight forward from the first, at the same height from the floor of the cabin: will the light leave the cabin through that hole? (c) How many meters will the pulse of light fall vertically? Suppose the width of the cabin is 2 m.

Solution 2. Like before rewrite the equations 9.28 in such a manner that the σ_y axis is vertically oriented upward and the σ_x axis is horizontally oriented parallel with respect to the spatial velocity of the light pulse. Axes' common origins coincide with the first hole, from which the light pulse enters the cabin. The equation of motion is therefore:

$$\begin{cases} \sigma_x(t) = v_{ox}t \\ \sigma_y(t) = \frac{c^2}{A}\left(\sqrt{1+\left(\frac{At+v_{oy}\gamma(v_{oy})}{c}\right)^2} - \gamma(v_{oy})\right) \end{cases} \quad (9.31)$$

where $v_{oy} = 0\,\text{m/s}$, since the light pulse is shot horizontally, $v_{ox} = c$ and, by invoking the *SEP*, the magnitude of the 4−acceleration is $A = -g$, where the minus derives from the fact that the *y* axis and $\vec{g}$ are anti-parallel. So:

$$\begin{cases} \sigma_x(t) = ct \\ \sigma_y(t) = -\frac{c^2}{g}\left(\sqrt{1+\left(\frac{gt}{c}\right)^2} - 1\right). \end{cases} \quad (9.32)$$

These constitute the parametric equations of an equilateral hyperbola, as showed in Figure 9.1.

In fact, as usual, by substituting $t = \sigma_x/c$ in the second equation it yields:

$$\sigma_y(\sigma_x) = -\frac{c^2}{g}\left(\sqrt{1+\left(\frac{g\sigma_x}{c^2}\right)^2} - 1\right) \quad (9.33)$$

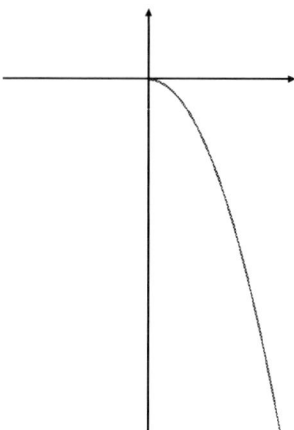

Figure 9.1.

in which, being σ_x positive and σ_y negative[10], is equivalent to the part of the equilateral hyperbola 9.34 situated in the fourth-quadrant:

$$\left(\frac{\sigma_x}{\frac{c^2}{g}}\right)^2 - \left(\frac{\sigma_y - \frac{c^2}{g}}{\frac{c^2}{g}}\right)^2 = -1. \tag{9.34}$$

Standing the fact that the speed acquired by the cabin during the very small interval of time necessary to the pulse of light to reach the opposite wall of the cabin is surely much smaller than c, taking into account the classical approximate relationships 9.6.,

$$\begin{cases} \sigma_y(t) \approx -\tfrac{1}{2}gt^2 + 0t + 0 = -\tfrac{1}{2}gt^2 \\ \sigma_x(t) = ct = 2\,\text{m} \end{cases} \rightsquigarrow \begin{cases} t = \tfrac{2}{c} \approx 6.7 \cdot 10^{-9}\,\text{s} \\ \sigma_y\left(\tfrac{2}{c}\right) \approx -\tfrac{1}{2}g\tfrac{4}{c^2} \approx -2.2 \cdot 10^{-16}\,\text{m}. \end{cases}$$

On the other hand, by invoking WEP$_c$, also the cabin - and so the right hole - fall vertically of $2.2 \cdot 10^{-16}$ m so that the light will leave the cabin through the second hole.

Note that in order to answer to (b) it was sufficient to invoke SEP$_{\mathcal{E}}$. In fact, the elevator is in free fall, so, according to *Einstein Equivalence Principle*, the laws of physics in the immediate neighbourhood of the cabin are

---

[10] The values of σ_x are positive since the pulse of light starts its motion from $\sigma_x = 0$ towards the increasing values of σ_x; the values of σ_y are only negative since in (9.33) the radicand is always greater or equal 1.

The Accelerated Motion

those of special relativity: there is no gravity so light travels at a constant speed along straight lines. Therefore, light entering the cabin horizontally will continue to travel in the horizontal direction. Since there is a second hole in the elevator wall, straight toward the first, the light will leave the cabin through that hole. □

Problem 3 (Curvature of a ray of light). *Suppose to be in a region of the Universe where the gravity is uniform with magnitude g. A pulse of light is shot with angle θ (elevation or depression) with respect to a fixed plane perpendicular to the gravitational field. Compute the maximum curvature of the trajectory of the pulse of light.*

Solution 3. Remember that the maximum curvature of the hyperbola's upper branch $\sigma_y = b\sqrt{1 + \left(\frac{\sigma_x}{a}\right)^2}$ takes place in $\sigma_x = 0$ and its value is $k(0) = b/a^2$. So the maximum curvature of the hyperbola 9.30 realizes in $\sigma_x = 0$ and its value is

$$k(0) = \frac{c^2}{A}\left(\frac{cv_{ox}}{A}\right)^{-2} = \frac{A}{v_{ox}^2} = \frac{g}{c^2 \cos^2\theta}.$$

Note that for every fixed value of $|\theta|$ (remember that θ denotes the elevation or the depression angle) since the square of the cosine is an even function, the curvature doesn't change. Besides, since translations is an isometry and the curvature is an intrinsic property of a curve, despite the maximum of the curvature of the hyperbola 9.29 realizes in a different point, namely $\sigma_x = -\frac{v_{oy}\gamma(v_{oy})}{c}$, its value is anyway $\frac{g}{c^2 \cos^2\theta}$.

Suppose to light horizontally the front wall. Then the maximum curvature of the trajectory of the light emitted is $g/c^2 \approx 1.1 \cdot 10^{-16}$ m^{-1}, in other words, the minimum ray of the osculating circle is approximately equal to 10^{16} m $\approx$ 1 ly and the trajectory is (substantially) a straight line. □

9.8. Speed versus Time Graphs

Previously, in equation (9.25) the functions $\tau = \tau(t)$ were deduced in uniformly relativistic accelerated motion. Besides, remember that, as stated in section 8.5., the coordinate "σ" denotes a proper length (with sign) while t is a non-proper time.

So, for each instant t, $v(t) = \sigma(t)/t$ given by (9.8) is the equation for the speed, whose graph with respect to t is clearly neither a straight line (as in Newtonian's uniformly accelerated motion) nor an hyperbola but a curve which tends asymptotically to $v = c$.

What about the analogous graph for the speed determined as the quotient of a *non-proper* length, "s", measured by someone or something whose clock beats the proper time τ? According to the equation $v = v(\tau)$ in (9.21) the graph of v versus the proper time τ is evidently *not* the same as $v = v(t)$. In any case, once again, it is neither a straight line nor an hyperbola but a curve that tends asymptotically to $v = c$.

9.9. Rindler's Metric for L-M Spacetime

Defining Rindler's Coordinates and Metric

Remember the equations 9.18 which, if $v_0 = 0$, by switching the notation with "x" instead of "s" (the non-proper spatial coordinate) looks like this:

$$\begin{cases} t(\tau) = \frac{c}{A} \sinh\left(\frac{A}{c}\tau\right) \\ x(\tau) = \frac{c^2}{A} \left(\cosh\left(\frac{A}{c}\tau\right) - 1\right). \end{cases} \quad (9.35)$$

Standing these two equations, it's quite natural to consider the following change of coordinates:

$$\begin{cases} x(\xi, \eta) = \xi \cosh(\eta) \\ t(\xi, \eta) = \frac{\xi}{c} \sinh(\eta). \end{cases} \quad (9.36)$$

The reason for this is a true changing of coordinates is that it's possible to exhibit its inverse transformation. To achieve this purpose, after having homogenized the spaces in times, applying the fundamental hyperbolic identity it easily follows that $(x/c)^2 - t^2 = (\xi/c)^2$ so that it must be $|x/c| \geq |t|$ and, if $\xi \geq 0$, $\xi = \sqrt{x^2 - (ct)^2}$. Besides, referring again to (9.36), by computing t/x it results

$$\eta = \operatorname{artanh}\left(\frac{t}{x/c}\right) = \frac{1}{2} \ln \frac{x/c + t}{x/c - t}$$

so that

$$\xi > 0 \Rightarrow \begin{cases} \xi(t,x) = \sqrt{x^2 - (ct)^2} \\ \eta(t,x) = \frac{1}{2}\ln\frac{x+ct}{x-ct} \end{cases} \quad (9.37)$$

$$\xi < 0 \Rightarrow \begin{cases} \xi(t,x) = -\sqrt{x^2 - (ct)^2} \\ \eta(t,x) = \frac{1}{2}\ln\frac{x+ct}{x-ct}. \end{cases} \quad (9.38)$$

In particular both (9.37) and (9.38) constitute a local inverse transformation of (9.36) provided $\xi \neq 0$ ($\xi = 0$ implies $x = ct$ so that η is no longer defined) if and only if $\left|\frac{x}{c}\right| > |t|$. Hence

$$(\xi,\eta): \{(x,t) \in \mathbb{R}^2 : |x/c| > |t|\} \longmapsto (\xi(x,t), \eta(x,t)) \in \{\mathbb{R}\setminus\{0\}\} \times \mathbb{R}$$

given by (9.37) and (9.38), being differentiable with its inverse (9.36) constitute a *changing of coordinates*. The coordinates (ξ,η) constitute themselves two disjointed charts for the Lorentz-Minkowsky space, namely (9.37) and (9.38) do not overlap each other. Besides they do not cover the entire space, as a matter of fact, ξ can't be equal to zero, which means that the two charts do not cover the entire line corresponding to $\xi = 0$. **These two systems of coordinates can be used only on each of the two half-planes singularly.** It will be possible to come back to this later (see § 9.9. point 5).

The system (ξ,η) are said to be *Rindler's coordinates* for the Lorentz-Minkowsky space and their corresponding metric *Rindler's metric*.

Let's compute Rindler's metric: first of all, starting from (9.36) it follows immediately that

$$\begin{cases} dt = \frac{\partial t}{\partial \xi}d\xi + \frac{\partial t}{\partial \eta}d\eta = \frac{1}{c}\sinh(\eta)d\xi + \frac{\xi}{c}\cosh(\eta)d\eta \\ dx = \frac{\partial x}{\partial \xi}d\xi + \frac{\partial x}{\partial \eta}d\eta = \cosh(\eta)d\xi + \xi\sinh(\eta)d\eta \end{cases} \quad (9.39)$$

so

$$d\tau^2 = dt^2 - \left(\frac{dx}{c}\right)^2 = \frac{1}{c^2}\sinh^2(\eta)d\xi^2 + \frac{\xi^2}{c^2}\cosh^2(\eta)d\eta^2 + 2\frac{\xi}{c^2}\sinh(\eta)\cosh(\eta)d\xi d\eta$$

$$-\frac{1}{c^2}\cosh^2(\eta)d\xi^2 - \frac{\xi^2}{c^2}\sinh^2(\eta)d\eta^2 - 2\frac{\xi}{c^2}\cosh(\eta)\sinh(\eta)d\xi d\eta = \left(\frac{\xi}{c}\right)^2 d\eta^2 - \frac{1}{c^2}d\xi^2$$

that is

$$\boxed{d\tau^2 = \left(\frac{\xi}{c}\right)^2 d\eta^2 - \frac{1}{c^2}d\xi^2} \quad (9.40)$$

Some observations are needed.

Metric Interpretation

It follows an useful analysis of the Rindler's metric. In particular these three circumstances: $d\xi = 0$, $d\eta = 0$ and $d\tau = 0$ will be deeply analyzed.

1. Since (9.40) is diagonalized, given that is adopted from the beginning the *West Coast Convention*, it follows immediately that η is the *temporal* coordinate while ξ is the *spatial* one.

2. Let the spatial coordinate ξ be fixed, so (9.40) becomes $d\tau^2 = \left(\frac{\xi}{c}\right)^2 d\eta^2$ iff $d\tau = \left|\frac{\xi}{c}\right| d\eta$, so, by requiring ($\eta = 0 \Rightarrow \tau = 0$) an integration yields $\tau = \left|\frac{\xi}{c}\right| \eta$. To fix ideas suppose now $\xi > 0$ so that $\tau = \frac{\xi}{c}\eta \Leftrightarrow \eta = \frac{c\tau}{\xi}$. With this substitution equations 9.36 become

$$\begin{cases} t(\xi, \eta) = \frac{\xi}{c} \sinh\left(\frac{c}{\xi}\tau\right) \\ x(\xi, \eta) = \xi \cosh\left(\frac{c}{\xi}\tau\right). \end{cases} \quad (9.41)$$

First of all, note the strong relationship between these new equations and (9.35). From a direct comparison, it is possible to immediately deduce that $\frac{c^2}{\xi}$ coincides with the uniform acceleration A. On the other hand, as already noticed in point 1, ξ is a spatial variable.

Here it is possible to say to know everything about 9.41. It is possible then to say with certainty to have obtained (9.36) and so (9.41) starting from (9.18) which descend themselves from the equations of relativistic accelerated motion, which are deeply analyzed in lots of the previous sections. What about if (9.36) had been written first? Surely it needs to have a certain insight, equations (9.36) are all but not so obvious to be written stand-alone! Surely, starting from (9.36) have probably saved many difficult computations. For example, it could have been possible to deduce that they represent the equations of a uniformly accelerated motion by computing the 4−derivative with respect to proper time, discovering that it is constant, just $\frac{c^2}{\xi}$! Anyway, equations 9.41 have still something to tell us. Some very interesting consequences of it will be analyzed in the following § 9.9.

3. As further specification regarding point 2 above, it is necessary to underline the fact that in Rindler's coordinates the time-coordinate

The Accelerated Motion

η does not measure the time of a clock whose spatial coordinate ξ is fixed, i.e. the proper time. On the contrary, considering the Lorentz-Minkowsky metric

$$d\tau^2 = dt^2 - \left(\frac{dx}{c}\right)^2,$$

if the spatial coordinate x is fixed than $d\tau = dt$, and so $\tau = t$ (if $t = 0 \Rightarrow \tau = 0$). In Rindler's metric the corrective factor $\left|\frac{\xi}{c}\right|$ is needed.

4. Let the temporal coordinate η be fixed, so (9.40) becomes

$$d\tau^2 = -\left(\frac{d\xi}{c}\right)^2 \Rightarrow d\tau = \sqrt{-d\tau^2} = \frac{d\xi}{c}$$

where ξ is assumed to be oriented towards the increasing values of spaces. Besides, the distance between two events having the same temporal coordinate η is merely the difference between their spatial coordinates, i.e. $|\xi_2 - \xi_1|$. This can be immediately proven by integrating $d\Sigma = -cd\tau = d\xi$.

5. Let the proper time τ be fixed, so that $d\tau = 0$. This corresponds to looking for the geodesic of light, i.e. those lines along which time never flows. In order to achieve this target, starting from (9.40), it becomes

$$\left(\frac{\xi}{c}\right)^2 d\eta^2 = \left(\frac{d\xi}{c}\right)^2 \Leftrightarrow \xi d\eta = \pm d\xi \Rightarrow \eta = \pm \ln\left|\frac{\xi}{k}\right| = \pm \, \ln|\xi| - \ln|k|)$$

where the constant k belongs to $\mathbb{R}_0$. This solution consists of two classes of curves, the first one with the *plus sign*, the second one with *the minus*. Fixed one class of curves, each curve differ from another for a translation directed along the η axis.

Last but not the least, ξ and k agree in sign. In particular, ξ can't be zero[11], which physically means that the two sub-manifolds corresponding to $\xi > 0$ and $\xi < 0$ do not communicate with each other: **no light signal starting from a point with $\xi > 0$ can reach another point with $\xi < 0$, and vice-versa.**

---
[11] As it is immediately possible to see, the metric is singular for $\xi = 0$, in fact, its determinant equals to $-\xi^2/c^2$.

In conclusion, it is possible to say that the Lorentz-Minkowsky space with Rindler's metric is *viewed from a uniformly accelerated frame*.

Time Dilation in Rindler's Metric

With reference to Figure n. , in $A = (\xi_A, \eta_A)$ consider someone or something which send a light signal arriving in $B = (\xi_B, \eta_B)$. So, consider another light signal starting from $A' = (\xi_A, \eta_{A'})$ which arrives in $B' = (\xi_B, \eta_{B'})$. Surely

$$\Delta\eta := \eta_{A'} - \eta_A = \eta_{B'} - \eta_B$$

since the geodesics of light differ each other for a translation (see point n. 5 above) which is a particular isometry.

Now, taking into account the point 2 above, it results that $\Delta\tau_1 := \frac{\xi_1}{c}\Delta\eta$ and $\Delta\tau_2 := \frac{\xi_2}{c}\Delta\eta$ so that

$$\frac{\Delta\tau_2}{\Delta\tau_1} = \frac{\xi_2}{\xi_1} \Rightarrow \Delta\tau_2 = \frac{\xi_2}{\xi_1}\Delta\tau_1 = \frac{\xi_1 + \xi_2 - \xi_1}{\xi_1}\Delta\tau_1 = \left(1 + \frac{\xi_2 - \xi_1}{\xi_1}\right)\Delta\tau_1$$

$$= \left(1 + \frac{\Delta\xi}{\xi_1}\right)\Delta\tau_1 = \left(1 + \frac{c^2}{\xi_1}\frac{\Delta\xi}{c^2}\right)\Delta\tau_1 \quad (9.42)$$

on the other hand

$$\Delta\tau_1 = \frac{\xi_1}{\xi_2}\Delta\tau_2 = \left(1 + \frac{\xi_1 - \xi_2}{\xi_2}\right)\Delta\tau_2$$

$$= \left(1 + \frac{\Delta\xi}{\xi_2}\right)\Delta\tau_2 = \left(1 + \frac{c^2}{\xi_2}\frac{\Delta\xi}{c^2}\right)\Delta\tau_2 \quad (9.43)$$

where, as noted in point 2, $c^2/\xi_{1,2}$ is the magnitude of the 4−acceleration "A", while $\Delta\xi$ is the variation of the spatial coordinate, which can be thought, for example, as a variation of altitude. Note that the equations 9.42 and analogously 9.43 have been derived under the assumption that the magnitude of the acceleration $A = c^2/\xi_{1,2}$ would be constant and the spatial acceleration $\vec{a}$ would be uniform in direction, namely under the assumption of *uniformly relativistic accelerated motion*. So, in order to respect the interpretation of the Rindler's metric as in § 9.9., ξ_1 need to be approximately equal to ξ_2 so that $c^2/\xi_1 \approx c^2/\xi_2$; besides the direction of the scalar acceleration $\vec{a}$ need to be uniform.

The Accelerated Motion 167

To better understand the above-prospected situation consider the following:

Problem 4. *Consider two identical clocks in two different places whose difference in altitude consists in* 3250 m. *Clock one (lower than clock two) emits an initial light-type signal of initial synchronism and after* 68 *days it emits a final signal of final synchronism, ending the experiment. Compute the difference in time between the two clocks.*

Solution 4. The variation in spatial coordinate between the two clocks is $\Delta\xi = \Delta h = 3250$ m. On the other hand, invoking the SEP, the magnitude of the relativistic acceleration $A(\tau,s) = a(\tau)$ which, in this case, can be thought a constant function of proper time τ. As a matter of fact, $\vec{a}(\tau)$ can be substituted with the gravity acceleration $\vec{g}$. In this respect note that the gravity acceleration does not undergo a sensible variation due to the slope of 3250 m (as a matter of fact the variation is just over 1%). Regarding the presumed difference in the direction of the two gravity vectors, suppose that the two clocks lie in two different places whose geographic coordinates are approximately the same so that there isn't even a remarkable difference in direction. Besides note that, standing the time intervals below specified and being $v_0 = 0$ it results that $a(\tau) \approx A = A(t,\sigma)$, so even A is approximately constant.

Standing the premises above it is possible to apply the above prospected relationships. Let $\Delta\tau_1 = 68\,\text{days} = 5\,875\,200.000\,0000$ s, so that

$$\Delta\tau_2 = \left(1 + \frac{g\Delta h}{c^2}\right)\Delta\tau_1 = \left(1 + \frac{9.81 \cdot 3250}{299792458^2}\right) 5\,875\,200.000\,0000\,\text{s}$$

$$\approx 5\,875\,200.000\,0021\,\text{s}$$

so $\Delta\tau_2 - \Delta\tau_1 = 5\,875\,200.000\,0021\,\text{s} - 5\,875\,200.000\,0000\,\text{s} = 2.1\,\mu\text{s}$. □

Later in chapter 11 the same problem will be solved by using another metric, namely the so famous Schwarzschild's metric, which is not deducible with a change of coordinate from the Lorentz-Minkowsky metric, as Rindler's metric does. Surprisingly the identical result will be found, up to the seventh decimal figure.

As a matter of fact, the Schwarzschild's spacetime endowed with his metric, *is not flat* - as the Lorentz-Minkowsky space is - since the Christoffel symbols of the metric (another invariant which depends only from the metric and not from the local chart used to describe locally the spacetime, see § 11.1. et seq.) are not all identically equal to zero.

9.10. Horizons

This section deals with three problems regarding accelerated motion. All of them are studied in a frame of reference *external* to the accelerating bodies. Further in this section, these problems will be deeply analyzed by using a different frame, coincident with the accelerating bodies themselves. Rindler's coordinates will turn out to be useful in order to achieve this target.

Problem 5 (Interstellar communications - part one). *The scout-ship "Discovery" assigned to explore the intergalactic space is moving away from the interstellar outpost "Deep Space Station One". Being Discovery an old generation spaceship, its guidance system needs to be updated by the station, which sends it a radio signal when needed. Suppose that Discovery starts its journey at rest with respect to the hangers, which are situated* 10 km *far from the space station, with uniform* 4−*acceleration quantified in the comfortable* 10 m/s². *According to the station's frame, the flight plan expects Discovery to uniformly accelerate for the first* 100 pc *towards M45. (a) In the frame of reference of the station, is there any instant* $\tilde{t}$ *such that for every* $t \geq \tilde{t}$ *the radio signal sent at time t will never reach the spaceship? (b) What time should the station send the radio signal so that Discovery will be able to receive it at the exact instant of the engine shutting down? (c) In the frame of reference of the spaceship, what distance will the spaceship cover at the instant of shutting down? (d) What time does the spaceship receive the signal sent by the station at time* $\bar{t}$? *Calculate this time both in the frame of the station, say* t_{rec} *and in the frame of the spaceship,* τ_{rec} *obtaining* $\tau_{rec} = \tau(t_{rec})$.

Solution 5. Fix a one dimensional coordinate system σ with its origin coincident with the hanger, parallel to the direction of motion of the spaceship. In the frame of reference of the Station, standing the above mentioned 1−dimensional coordinate system, the laws of motion of both the space

The Accelerated Motion

station and the radio signals are the following:

$$\text{station:} \quad \sigma(t) \equiv -\sigma_0 \tag{9.44}$$

$$\text{radio signal at time } \bar{t}: \quad \sigma(t) = c(t - \bar{t}) - \sigma_0 \tag{9.45}$$

where $\sigma_0 = 10\,\text{km}$ denotes the proper distance between the space station and the hanger. Besides, according to (9.10) the law of motion of the spaceship is

$$\text{spaceship:} \quad \sigma(t) = \frac{c^2}{A}\sqrt{1 + \left(\frac{At}{c}\right)^2} - \frac{c^2}{A}, \tag{9.46}$$

in which $A = 10\,\text{m/s}^2$. In particular, the worldline of the spaceship is represented in the $(\sigma/c, t)$ plane by the upper semi-branch lying in first quadrant of the following equilateral hyperbola:

$$\text{spaceship:} \quad \frac{\left(\frac{\sigma}{c} - \left(-\frac{c}{A}\right)\right)^2}{\left(\frac{c}{A}\right)^2} - \frac{t^2}{\left(\frac{c}{A}\right)^2} = 1.$$

The worldline of the radio signal transmitted by the space-station at time $\bar{t}$ is indeed

$$\text{radio signal:} \quad \left(\forall \bar{t} \in \mathbb{R}^+\right) \quad t = 1 \cdot \left(\frac{\sigma}{c}\right) + \frac{\sigma_0}{c} + \bar{t}.$$

Note that the whole of such worldlines constitute a bundle of straight lines parallel to the bisector $t = \frac{\sigma}{c}$, each of them being determined by exactly one $\bar{t} \in \mathbb{R}^+$.

At last, the worldline of the Station is merely given by

$$\text{station:} \quad \frac{\sigma}{c} \equiv -\frac{\sigma_0}{c}$$

that is a vertical straight line through the point $\left(-\frac{\sigma_0}{c}, 0\right)$.

The radio signal originated at time $\bar{t}$ reaches the spaceship if and only if their worldlines intersect each other, that is, the following system is determined:

$$\begin{cases} t = 1 \cdot \left(\frac{\sigma}{c}\right) + \frac{\sigma_0}{c} + \bar{t} \\ \dfrac{\left(\frac{\sigma}{c} - \left(-\frac{c}{A}\right)\right)^2}{\left(\frac{c}{A}\right)^2} - \dfrac{t^2}{\left(\frac{c}{A}\right)^2} = 1, \quad \sigma \geq 0. \end{cases}$$

Now, a trivial geometric argument[12] shows that this system admits solutions if and only if $\bar{t} + \frac{\sigma_0}{c} < \frac{c}{A}$. In other words, if $\bar{t} \geq \frac{c}{A} - \frac{\sigma_0}{c}$ then the spaceship results unreachable from the Station, even by a ray of light (a radio signal): from the point of view of the Station, the spaceship falls under the *event horizon* of the station itself. In particular, if the distance σ_0 was zero - which physically correspond to the condition "hanger located exactly in the station, in the same position of the radio transmitter" - then the radio signal emitted by the station will be able to reach the spaceship until the time of emission $\bar{t}$ results in less than c/A.

Note that each point $\left(\frac{\tilde{\sigma}}{c}, \tilde{t}\right)$ of the spaceship's worldline is the vertex of a double-cone whose mathematical description is given by the inequality $|t - \tilde{t}| \geq \left|\frac{\sigma}{c} - \frac{\tilde{\sigma}}{c}\right|$. Each point of this double cone belongs to the space-light - like cone, that is the set of all points which are connectible with $\left(\frac{\tilde{\sigma}}{c}, \tilde{t}\right)$ by a signal whose speed is less or equal than c. As long as $\bar{t} < \frac{c}{A} - \frac{\sigma_0}{c}$ the Universe line of the ray of light coming from the Station enters the light-cone sliding to its edge.

(a) In the frame of reference of the station, is there any instant $\tilde{t}$ such that for every $t \geq \tilde{t}$ the radio signal sent at time t will never reach the spaceship? Starting from $\bar{t} = \frac{c}{A} - \frac{\sigma_0}{c}$ the worldline of the ray of light originated at $\bar{t} = \tilde{t}$ exits definitively from every light-cone whose vertexes belongs to the spaceship's worldline. In details it results:

$$\tilde{t} = \frac{c}{A} - \frac{\sigma_0}{c} \approx \frac{c}{A} \approx 29\,979\,245.8\,\text{s} \approx 0.95\,\text{y}.$$

What time[13] "t" will the spaceship receive the signal emitted by the station at $\bar{t}$, namely the signal after which no other signal will be received by the spaceship? Standing the fact that for $t = \bar{t}$ the worldline of the signal emitted by the station coincides the oblique asymptote of the spaceship's worldline, they will intersect at $t = +\infty$, namely[14] the signal emitted by the station at $t = \bar{t} - \varepsilon$ ($0 < \varepsilon \ll 1$) will reach the spaceship when $t \to +\infty$, at a distance $\sigma \to +\infty$ from the space station.

---

[12]The asymptote of the hyperbola is $t = 1 \cdot \frac{\sigma}{c} + \frac{c}{A}$ so that $t = 1 \cdot \left(\frac{\sigma}{c}\right) + \frac{\sigma_0}{c} + \bar{t}$ intersects the branch of hyperbola as long as $\frac{\sigma_0}{c} + \bar{t} = \frac{c}{A}$.

[13]Remember that t denotes the time measured by the station's atomic clock.

[14]By an argument of continuity.

The Accelerated Motion

(b) What time should the station send the radio signal so that Discovery will be able to receive it at the exact instant of engine shutting down? Applying the equation 9.46 with $\sigma(t) = \sigma_{\text{fin}} = 100\,\text{pc} \approx 326\,\text{ly} \approx 3.08 \cdot 10^{18}$ m and solving it with respect to t yields:

$$t = \frac{c}{A}\sqrt{\left(\frac{A}{c^2}\sigma_{\text{fin}} + 1\right)^2 - 1} \approx 326.5\,\text{y}.$$

So, substituting this value in the law of motion of the radio signal together with $\sigma = \sigma_{\text{fin}}$ and solving it with respect to $\bar{t}$ yields:

$$\bar{t} = t - \frac{\sigma_{\text{fin}}}{c} - \frac{\sigma_0}{c} = 326.5\,\text{y} - \frac{326\,\text{ly}}{c} - \frac{10^3}{c} \approx 326.5\,\text{y} - 326\,\text{y} = 0.5\,\text{y}.$$

How far is the horizon of the station to which it can address radio signals? In other words, what distance will the spaceship be from the space station at the actual moment in which the last signal reachable from the spaceship will be emitted? According to the law 9.46 of uniform accelerated motion of the spaceship such a distance is:

$$\sigma(\bar{t}) = \frac{c^2}{A}\sqrt{1 + \left(\frac{A\bar{t}}{c}\right)^2} - \frac{c^2}{A} + \sigma_0 = \frac{c^2}{A}\left(\sqrt{1 + \left(1 - \sigma_0\frac{A}{c}\right)^2} - 1\right) + \sigma_0 \approx 0.39\,\text{ly}.$$

If $\sigma_0 = 0$ then the distance of the horizon of the events from the station is:

$$\sigma(\bar{t}) = \frac{(\sqrt{2}-1)c^2}{A}.$$

(c) In the frame of reference of the spaceship, what distance will the spaceship cover at the instant of shutting down? Equation 9.27 need to be inverted:

$$s = \frac{c^2}{A}\ln\left(\frac{A}{c^2}\sigma + 1\right) \approx 5.56\,\text{yl} \approx 1.7\,\text{pc},$$

another evidence of length contractions.

(d) What time does the spaceship receive the signal sent by the station at time $\bar{t}$? Calculate this time both in the frame of the station, say t_{rec} and in the frame of the spaceship, τ_{rec} obtaining $\tau_{\text{rec}} = \tau(t_{\text{rec}})$. Considering for sake of simplicity $\sigma_0 = 0$, making a system between the laws of motion 9.45 and 9.46, assuming evidently $\bar{t} < \frac{c}{A}$ it yields

$$t_{\text{rec}} = \frac{\left(2 + \frac{A}{c}\bar{t}\right)}{2\left(1 - \frac{A}{c}\bar{t}\right)}\bar{t}$$

which tends to $+\infty$ as $\bar{t} \to \left(\frac{c}{A}\right)^-$. Besides:

$$\tau(t_{\text{rec}}) = \frac{c}{A} \operatorname{arsinh}\left(\frac{\left(2+\frac{A}{c}\bar{t}\right)}{2\left(1-\frac{A}{c}\bar{t}\right)} \frac{A}{c}\bar{t}\right).$$

Note that, since $\bar{t} < \frac{c}{A}$ it results that $1 - \frac{A}{c}\bar{t} > 0$, hence

$$\lim_{\bar{t} \to \left(\frac{c}{A}\right)^-} \tau(t_{\text{rec}}) = +\infty.$$

At last note that $\operatorname{arsinh}(x) = O\left(\frac{e^x}{2}\right)$ as $x \to +\infty$. So, in conclusion, **as $\bar{t}$ tends increases its value up to $\frac{c}{A}$, then the corresponding instant of receiving τ_{rec} increases exponentially.** An analogous situation will be found in problem 6. □

Before going further it is fundamental to fully understand the fact that the above prospected *event horizon* can be removed at any time: it is sufficient for the spaceship to shut down the engines zeroing its uniform relativistic acceleration. As a matter of fact, if $v(\tau_{\text{shut}})$ denotes the speed reached by spaceship in its frame contextually to the engines' shutdown, then the spaceship's worldline turns to be a straight line whose gradient is greater than 1, so that the worldline of the radio signal sooner or later will intersect the spaceship's worldline.

Besides, standing the fact that the spaceship's speed is at each (finite) instant strictly less than c, how is it possible for the radio signal (whose speed is always equal to c) to fail to reach the spaceship? Set for sake of simplicity $\sigma_0 = 0$, so, if $\bar{t} \geq c/A$ then the radio signal will always remain backward with respect to the spaceship: its disadvantage decreases up to reach its inferior value as $t \to +\infty$ (so that the spaceship reaches its limit value c) and $\bar{t}$ decreases up to reach $\tilde{t} = c/A$. More intuitively, if $\bar{t} < c/A$ then the radio signal reach the spaceship before it gains speed equal to c; if $\bar{t} = c/A$ then the radio signal reach the spaceship at the exact instant in which it gains speed equal to c, at last, if $\bar{t} > c/A$ then the spaceship gains speed approximately equal to c before the radio signal reaches the spaceship.

Problem 6 (Interstellar communication - part two). *A spaceship depart from an interstellar space station with uniform 4−acceleration A. According to the spaceship's atomic clock, after a time quantified in $\bar{\tau}$ seconds*

The Accelerated Motion

from the departure, the spaceship sends a radio signal addressed to the space-station. What time does the space-station will receive the radio signal according to its atomic clock?

Solution 6. Let's try to solve the problem in the frame of reference of the *spaceship*. Fixed a one dimensional coordinate system s with its origin coincident with the space-station, parallel to the direction of motion of the spaceship, the laws of motion are the following:

$$\text{spaceship:} \quad s(\tau) = \frac{c^2}{A} \ln\left(\cosh\left(\frac{A}{c}\tau\right)\right)$$

$$\text{radio signal at time } \bar{\tau}: \quad s(\tau) = -c(\tau - \bar{\tau}) + \frac{c^2}{A} \ln\left(\cosh\left(\frac{A}{c}\bar{\tau}\right)\right)$$

$$\text{station:} \quad s(\tau) \equiv 0.$$

In particular, the worldline of the spaceship in the $(s/c, \tau)$ plane is given by

$$\text{spaceship:} \quad \tau(s) = \frac{c}{A} \operatorname{arcosh}\left(e^{\frac{A}{c} \cdot \frac{s}{c}}\right).$$

Regarding the worldline of the *radio signal*, it is:

$$\text{radio signal at time } \bar{\tau}: \quad \tau(s) = -\frac{s}{c} + \frac{c}{A} \ln\left(\cosh\left(\frac{A}{c}\bar{\tau}\right)\right) + \bar{\tau}.$$

Finally, the worldline of the station is merely given by

$$\text{station:} \quad \frac{s}{c} \equiv 0.$$

The signal emitted by the spaceship at $\tau = \bar{\tau}$ will reach the space station at time τ_{rec} (*rec* stands for *receiving*) if and only if the worldlines of both the radio-signal and the station intersect each other, if and only if there exists $\tau_{\text{rec}} > \bar{\tau}$ such that the two corresponding worldlines equal:

$$\begin{cases} \tau_{\text{rec}} = -\frac{s}{c} + \frac{c}{A} \ln\left(\cosh\left(\frac{A}{c}\bar{\tau}\right)\right) + \bar{\tau} \\ \frac{s}{c} = 0 \end{cases} \Rightarrow \tau_{\text{rec}} = \frac{c}{A} \ln\left(\cosh\left(\frac{A}{c}\bar{\tau}\right)\right) + \bar{\tau}. \tag{9.47}$$

Note that if someone would like to deduce the non-proper time t_{rec} corresponding to τ_{rec} it isn't allowed to apply the relation

$$t(\tau) = \frac{c}{A} \sinh\left(\frac{A}{c}\tau + \operatorname{arsinh}\left(\frac{v_0 \gamma(v_0)}{c}\right)\right) - \frac{v_0 \gamma(v_0)}{A}$$

since this has been deduced for only uniformly accelerated motions: as a matter of fact, the radio signal moves from the spaceship to the station with *uniform motion* at constant speed c. The only thing it is possible to deduce from this argument is directly implied by (9.47):

$$\Delta \tau = \tau_{\text{rec}} - \bar{\tau} = \frac{c}{A} \ln \left(\cosh \left(\frac{A}{c} \bar{\tau} \right) \right).$$

On the other hand, it is interesting to find t_{rec}, namely the time of receiving the radio signal registered by the space station's atomic clock. So, let's analyze the same problem in the frame of reference the space station, with the same coordinate system. With this new premise, let $\bar{t}$ be the non-proper time corresponding to the proper time of emission $\bar{\tau}$, so that the laws of motion are the following:

$$\text{spaceship:} \quad \sigma(t) = \frac{c^2}{A} \sqrt{1 + \left(\frac{At}{c} \right)^2} - \frac{c^2}{A}$$

$$\text{radio signal at time } \bar{t}: \quad \sigma(t) = -c(t - \bar{t}) + \frac{c^2}{A} \sqrt{1 + \left(\frac{A\bar{t}}{c} \right)^2} - \frac{c^2}{A}$$

$$\text{station:} \quad \sigma(t) \equiv 0.$$

Note that the non-proper time $\bar{t}$ corresponding to the time of emission $\bar{\tau}$ is given now by the relation:

$$t(\tau) = \frac{c}{A} \sinh \left(\frac{A}{c} \tau + \text{arsinh} \left(\frac{v_0 \gamma(v_0)}{c} \right) \right) - \frac{v_0 \gamma(v_0)}{A}$$

where v_0 corresponds to the initial instant of the uniformly accelerated motion of the spaceship, namely $v_0 = 0$, so that:

$$\bar{t} = \bar{t}(\bar{\tau}) = \frac{c}{A} \sinh \left(\frac{A}{c} \bar{\tau} \right).$$

So, substituting in the law of motion of the radio signal it follows:

radio signal at time $\bar{t}$: $\quad \sigma(t) = -c(t-\bar{t}) + \dfrac{c^2}{A}\sqrt{1+\left(\dfrac{A\bar{t}}{c}\right)^2} - \dfrac{c^2}{A}$

$\quad = -c\left[t - \dfrac{c}{A}\sinh\left(\dfrac{A}{c}\bar{\tau}\right)\right] + \dfrac{c^2}{A}\sqrt{1+\sinh^2\left(\dfrac{A}{c}\bar{\tau}\right)} - \dfrac{c^2}{A}$

$\quad = -ct + \dfrac{c^2}{A}\sinh\left(\dfrac{A}{c}\bar{\tau}\right) + \dfrac{c^2}{A}\cosh\left(\dfrac{A}{c}\bar{\tau}\right) - \dfrac{c^2}{A}$

$\quad = -ct + \dfrac{c^2}{A}\left(e^{\frac{A}{c}\bar{\tau}} - 1\right), \; \forall t \geq \bar{t}.$

So, the radio signal at time $\bar{t}$ will reach the space station at time t_{rec} such that

$$-ct_{\text{rec}} + \dfrac{c^2}{A}\left(e^{\frac{A}{c}\bar{\tau}} - 1\right) = 0 \quad \Longleftrightarrow \quad \boxed{t_{\text{rec}} = \dfrac{c}{A}\left(e^{\frac{A}{c}\bar{\tau}} - 1\right)}$$

Note that as $\bar{\tau}$ increases linearly the corresponding instant of receiving t_{rec} increases exponentially. This constitutes another similarity between the situations prospected in the actual and previous problems. □

By direct comparison of the results of the two problems above it's possible to detect immediately a substantial *asymmetry* between the relativistic uniformly accelerating spaceship and the station, which is instead an inertial system. As a matter of fact, if a light signal is emitted by the spaceship, then, sooner or later it will reach the space station. On the contrary, if the light signal is emitted by the station then it will reach the spaceship if and only if, at the moment of emission, the spaceship is not already too far away. This looks paradoxical if it is not considered that, on the one hand, the spaceship begins its journey with an initial speed equal to zero and never will reach the maximum speed attainable, c, on the other hand, the light signal starts its run with a certain disadvantage but with the maximum speed attainable. Besides, remember that in the first case (the station which sends messages to the spaceship) the spaceship will receive the last signal when t tends to infinity. Now, the analogy between the situations of the two above prospected situations is more strong, the only difference remaining consists in the presence of the so-called *horizon* which seems to be a peculiarity of the inertial frame, the *station*, in problem 5 "Interstellar communication - part one".

Let's now refer again to problem 5, this time in the frame of the spaceship: is there any instant $\tilde{\tau}$ such that for every $\tau > \tilde{\tau}$ the radio signal sent by the station at time t corresponding to τ will never reach the spaceship? No, there isn't. Suppose for sake of simplicity $\sigma_0 = 0$ and fix a one-dimensional system of coordinates with its origin coincident with the spaceship, positive oriented towards the station. The laws of motions are the following:

spaceship: $\quad s(\tau) \equiv 0$

station: $\quad s(\tau) = \dfrac{c^2}{A} \ln\left(\cosh\left(\dfrac{A}{c}\tau\right)\right)$

signal starting at time $\bar{\tau}$: $\quad s(\tau) = -c(\tau - \bar{\tau}) + \dfrac{c^2}{A} \ln\left(\cosh\left(\dfrac{A}{c}\bar{\tau}\right)\right)$.

So the event *the signal arrives at the spaceship* realizes if and only if the following equations are determined:

$$-c(\tau - \bar{\tau}) + \dfrac{c^2}{A} \ln\left(\cosh\left(\dfrac{A}{c}\bar{\tau}\right)\right) = 0.$$

As a matter of fact, this equation is determined *for every value of* $\bar{\tau}$, which means that the corresponding worldlines of both the spaceship and the signal intersect each other for every value of $\bar{\tau}$.

This confirms the fact that the horizon above prospected is a peculiarity of the inertial frame *station*.

Horizon's Analysis in Rindler's Coordinates

Let's now analyze the situation prospected in problem 5 in Rindler's coordinates. In such a system of coordinates, what is the worldline of the uniformly accelerated spaceship-like? Remembering that $A = c^2/\xi$, in the (ξ, η) plane the spaceship's worldline is represented by a straight line parallel to the η axis: $\xi = c^2/A$, in other words it corresponds to the worldline of a stationary body in the $(\sigma/c, t)$ plane. And, what about the worldline of the space-station which is stationary in the $(\sigma/c, t)$ plane? Fixed a one-dimensional coordinates system with its origin coincident with hangers, directed from the spaceship to the station its results:

$$s(\xi, \eta) = x(\xi, \eta) = \xi \cosh\eta = s_0 \;\Rightarrow\; \eta = \text{arcosh}\left(\dfrac{s_0}{\xi}\right)$$

where $s_0 > 0$ denotes the coordinate of the station. In order to sketch the graph of the function $\eta = \eta(\xi)$ it suffices to take notice that each point belonging to the graph it must be enclosed inside the light cone. Besides it is known know that the η axis cannot be crossed.

It is possible to say that in Rindler's coordinates the situation prospected by problem 5 is reversed: the accelerated body, namely the spaceship, is "stationary" (since its worldline is a vertical straight line) while the fixed body, the station, is moving away. So the radio signals sent by the station to the spaceship corresponds to signals sent by a moving source towards a stationary target. In other words, it is possible to expect that the situation is the same as the one prospected in solution 6. Indeed it is just like something like this. As a matter of fact, consider in Rindler's coordinates the worldline of the light signal starting from the generic point belonging to the worldline of the station, say $R = \left(\widehat{\xi}, \text{arcosh}\left(\frac{s_0}{\widehat{\xi}}\right)\right)$. Its equation is:

$$\eta = \ln(\xi) + \text{arcosh}\left(\frac{s_0}{\widehat{\xi}}\right) - \ln\left(\widehat{\xi}\right).$$

Consider then the intersection of the worldline above with the worldline $\xi = \frac{c^2}{A}$ of the spaceship: it is:

$$S = \left(\frac{c^2}{A}, \ln\left(\frac{c^2}{A}\right) + \text{arcosh}\left(\frac{s_0}{\widehat{\xi}}\right) - \ln\left(\widehat{\xi}\right)\right).$$

Let's compare the variation of the coordinate η of S with the analogous coordinate of the starting point R belonging to the worldline of the station:

$$d\eta_R = \text{arcosh}'\left(\frac{s_0}{\widehat{\xi}}\right) d\widehat{\xi}$$

$$d\eta_S = \left(\text{arcosh}'\left(\frac{s_0}{\widehat{\xi}}\right) - \ln'(\widehat{\xi})\right) d\widehat{\xi} = d\eta_R - \frac{1}{\widehat{\xi}} d\widehat{\xi}$$

so that, remembering that the derivative of $x \mapsto \text{arcosh}\, x$ is $1/\sqrt{x^2 - 1}$,

$$\frac{d\eta_S}{d\eta_R} = 1 - \frac{1}{\widehat{\xi}} \frac{1}{\eta_R'} = 1 - \frac{1}{\widehat{\xi}} \sqrt{\left(\frac{s_0}{\widehat{\xi}}\right)^2 - 1} \frac{\widehat{\xi}^2}{-s_0} = 1 + \sqrt{1 - \left(\frac{\widehat{\xi}}{s_0}\right)^2}$$

which tends to 2 as the time-coordinate $\eta_R \to +\infty$ and so $\widehat{\xi} \to 0^+$. This means that, asymptotically, if the infinitesimal separation between two distinct radio signals coming from the station is considered, $d\eta_R$, then the corresponding arrivals in the spaceship are doubly separated, namely $d\eta_S = 2d\eta_R$. In detail, as η_R increases and correspondingly $\widehat{\xi}$ decreases (always remaining greater than 0), being

$$d\eta_S = \left(1 + \sqrt{1 - \left(\frac{\widehat{\xi}}{s_0}\right)^2}\right) d\eta_R$$

it results that $d\eta_S$ increases with respect to $d\eta_R$ up to become, asymptotically, twice $d\eta_R$. Physically it means that the signal received from the accelerated spaceship are more and more separated. Asymptotically this separation does not tend to ∞, on the contrary, as already discussed and demonstrated, it tends to be $d\eta_S = 2d\eta_R$.

In other words, it has been given proof of the fact that the situation prospected in problem 5 is the same as the one treated in problem 6, besides the two solutions are substantially the same, even in Rindler's coordinates. Though, standing the specificity of these systems of coordinates, as already observed, asymptotically the time separation do not tend to ∞ but tends to be twice: $d\eta_S = 2d\eta_R$.

Relative Relativistic Uniformly Accelerated Motions

Now it is going to be analyzed the relative motion of an accelerated frame with respect to another, namely the motion of a relativistic uniformly accelerated frame studied from another relativistic uniformly accelerated one. More in details, consider two spaceships, say SS_1 and SS_2 both uniformly accelerated with four-acceleration A_1 and A_2 respectively with respect to the same inertial frame. What about the motion of the first spaceship with respect to the second one? Is it still uniformly accelerated? First of all, if, as usual, σ_i denotes the proper distance of SS_i measured in a given system of coordinates of an inertial frame equipped with a clock that marks the time t, the relative distance between SS_1 and SS_2 is $\sigma_2 - \sigma_1$, which is positive if and only if SS_2 overtakes SS_1 and is negative otherwise. Besides, suppose as usual $\sigma_1(0) = 0 = \sigma_2(0)$ together with $v_1(0) = v_2(0)$, so that

The Accelerated Motion 179

both 9.27 and its inverse

$$s(\sigma) = \frac{c^2}{A} \ln\left(\frac{A}{c^2}\sigma + 1\right) \tag{9.48}$$

hold. Under the assumption that $A_2 > A_1 > 0$ so that SS_2 overtakes SS_1 at the starting grid, namely, for all $t > 0$ it results $\sigma_2 > \sigma_1$, the distance between SS_2 and SS_1 measured in the frame of SS_1 is given by

$$s_1(\sigma_2 - \sigma_1) = s_1\big(\sigma_2(t) - \sigma_1(t)\big) = s_1\big(\sigma_2(t(\tau_1)) - \sigma_1(t(\tau_1))\big)$$

where $\sigma_i = \sigma_i\big(t(\tau_1)\big) = \sigma_i(\tau_1) = \frac{c^2}{A_i}\left(\cosh\left(\frac{A_i}{c}\tau_1\right) - 1\right)$ so that, having denoted with τ_1 the time marked by the clock of SS_1:

$$s_1(\sigma_2 - \sigma_1) = \frac{c^2}{A_1}\ln\left\{\frac{A_1}{c^2}\left[\frac{c^2}{A_2}\left(\cosh\left(\frac{A_2}{c}\tau_1\right) - 1\right) - \frac{c^2}{A_1}\left(\cosh\left(\frac{A_1}{c}\tau_1\right) - 1\right)\right] + 1\right\}$$

or equivalently

$$s_1(\tau_1) = \frac{c^2}{A_1}\ln\left\{\frac{A_1}{c^2}\left[\frac{c^2}{A_2}\left(\cosh\left(\frac{A_2}{c}\tau_1\right) - 1\right) - \frac{c^2}{A_1}\left(\cosh\left(\frac{A_1}{c}\tau_1\right) - 1\right)\right] + 1\right\}. \tag{9.49}$$

Note that $\tau_1 = 0$ imply $s_1(0) = \frac{c^2}{A_1}\ln(1) = 0$, as desired.

In conclusion SS_2 viewed in the frame of SS_1 is neither uniformly accelerated with acceleration A_1 nor with acceleration A_2.

In details, since the law of motion 9.49 is not of (9.24) - type, namely $s(\tau) = \frac{c^2}{A}\ln\left(\cosh\left(\frac{A}{c}\tau\right)\right)$ for a fixed value of A, the motion of SS_2 in the frame of SS_1 **is not** uniformly accelerated. Note that even if $|A_1| = |A_2|$ the motion of SS_2 in the frame of SS_1 is *not* uniformly accelerated.

At last, under the above-mentioned assumptions, the domain of 9.49 consists in the set of all τ_1 such that

$$\frac{A_1}{c^2}\big(\sigma_2(\tau_1) - \sigma_1(\tau_1)\big) + 1 > 0 \iff \sigma_2(\tau_1) - \sigma_1(\tau_1) > -\frac{c^2}{A_1}.$$

On the other hand, standing the fact that for all $t > 0 \iff \tau_1 > 0$ it results that $\sigma_2 > \sigma_1$, being $A_1 > 0$, the existence's conditions of 9.49 are always satisfied.

9.11. Moving Away in Opposite Directions

Problem 7 (Interstellar communication - part three). *Two spaceships, say SS_1 and SS_2 depart from the same point in the Universe with opposite directions, far away from any massive bodies. Their motion is uniformly accelerated with four-accelerations A_1 and A_2 respectively, their initial velocities are both zero. The two spaceships SS_1 and SS_2 are both equipped with an atomic clock each of one beats its proper time, respectively τ_1 and τ_2. At the starting moment the two clocks agree so that $\tau_1 = \tau_2 = 0$. (a) Write down the law of motion of SS_2 in the frame of reference of SS_1. (b) According to SS_2's atomic clock, after a time quantified in $\bar\tau_2$ seconds from the departure, SS_2 sends a radio signal addressed to SS_1. Show that as τ_2 increases linearly the corresponding time of receiving $\tau_{1,rec}$ increases exponentially.*

Solution 7. The solution of this problem is very similar to the analogous one of problem 6 in which will however be taken into account the relative relativistic acceleration previously studied.

In the following solution, SS_2 could be thought of as a star that is irradiating in all directions and so, in particular, towards SS_1, which can be idealized itself as the planetary spaceship constituted by the entire planet Earth.

With these premises, after having taken a 1−dimensional system of coordinates with its origin in SS_1, anti-parallel with respect to SS_2, the law of motion of SS_2, say the faraway star, written in the frame of SS_1, in which itself is stationary, is:

$$s_1(\tau_1) = -\frac{c^2}{|A_1|} \ln\left\{\frac{|A_1|}{c^2}\left[\frac{c^2}{|A_2|}\left(\cosh\left(\frac{|A_2|}{c}\tau_1\right) - 1\right) + \frac{c^2}{|A_1|}\left(\cosh\left(\frac{|A_1|}{c}\tau_1\right) - 1\right)\right] + 1\right\}$$

$$= -\frac{c^2}{|A_1|} \ln\left[\cosh\left(\frac{|A_1|}{c}\tau_1\right) + \frac{|A_1|}{|A_2|}\left(\cosh\left(\frac{|A_2|}{c}\tau_1\right) - 1\right)\right]. \quad (9.50)$$

The law of motion of the radio signal emitted by SS_2 at time $\bar\tau_1$ is:

$$s_{1,\text{radio}}(\tau_1) = +c(\tau_1 - \bar\tau_1) + s_1(\bar\tau_1).$$

At last, the law of motion of SS_1, the Earth is:

$$s_1(\tau_1) \equiv 0.$$

In order to answer (b) it remains to transform $\bar\tau_1$ in the frame of SS_2 and then solving $s_{1,\text{radio}}(\tau_1) = 0$. Since this looks not easy at all, standing

The Accelerated Motion   181

the fact that the relative motion of SS_2 with respect to SS_1 is not *uniformly* accelerated so that it is not known the law of transformations of times, let's make another consideration. Standing the fact that the *jerk* term $\frac{|A_1|}{|A_2|}\left(\cosh\left(\frac{|A_2|}{c}\tau_1\right)-1\right)$ is an increasing monotonic positive function of τ_1, being the logarithm itself an increasing monotonic function, the space expressed by (9.50) will increase faster than if there were no jerk term, namely will increase faster then in the relativistic *uniformly* accelerated motion with uniform acceleration $|A_1|$. In other terms, the modulus of the acceleration increases as time τ_1 passes. In order to produce the expected result, it is possible to approximate the accelerated motion by studying it insufficiently small time intervals, in which of them the 4−acceleration could be considered uniform so that in the next interval the constant 4−acceleration is bigger than in the actual one. Following this idea, it is possible to apply to each interval the solution of problem 6 obtaining that, globally, as $\bar{\tau}_2$ increases linearly, the corresponding instance of receiving $\tau_{1,\text{rec}}$ increases exponentially, in which the coefficient of $\bar{\tau}_2$ is suitable.

In details, having determined a suitable partition $\bigcup_{i\in\mathbb{N}}[\tau_{1,i},\tau_{1,i+1}[$ of $\mathbb{R}^+$, for every $\tau_1 \geq 0$ there exists only one $j \in \mathbb{N}$ such that $\tau_1 \in [\tau_{1,j},\tau_{1,j+1}[$ so that

$$s_1(\tau_1) \simeq -\frac{c^2}{|A_1|+\delta_j}\ln\left[\cosh\left(\frac{|A_1|+\delta_j}{c}\tau_1\right)\right], \qquad (9.51)$$

where δ_j denotes an appropriate function constant over $[\tau_{1,j},\tau_{1,j+1}[$ for which the above approximate equation holds. Now, for every fixed value of $\bar{\tau}_2$ there exists $j \in \mathbb{N}$ such that $\tau_{1,\text{rec}} \in [\tau_{1,j},\tau_{1,j+1}[$. So (9.51) holds, hence - in virtue of problem 6's solution - it results that

$$\tau_{1,\text{rec}} = \frac{c}{|A_1|+\delta_j}\left(e^{\frac{|A_1|+\delta_j}{c}\bar{\tau}_2}-1\right),$$

as desired.

In view of this solution there is a substantial impossibility to communicate an information from SS_2 to SS_1 standing the fact that a pause in the communication for SS_2 consisting in $\bar{\tau}_2$ seconds correspond for SS_1 in a pause quantified in $\tau_{1,\text{rec}}$ seconds, which are possibly thousands of years for SS_1. □

In order to complete this long discussion, if the earliest formalisms of General Relativity are already known, the reader will be able to read immediately § 11.4.

9.12. A Little Excursion in Cosmology

Cosmology, from the Greek *Κοσμολογια*, namely *logos* around the order, the world, is concerned with the description of the origin, evolution and eventual fate of the Universe by physical laws.

Universe's Shape

The *Singularity - Single Point* Misconception

It is a common misconception that the Universe was originally a single point. It is crucial to correct that misconception.

There is a fixed number of points in space at any given cosmic time. As the Universe expands, distances between any two points become larger and larger. Extrapolating this back in time, it is found that the distances between any two points shrink. Continuing this process it is found that going further and further back in the past the distance between any two points tends to zero and the Universe tends to the *singularity* of the Big Bang.

Now, it's important to understand that in this process, points are not created or destroyed. Only the distances between them change. So even if there really was a singularity (which is a matter of debate) it's incorrect to say that *the Universe started as a single point and then expanded*. The correct statement is: *the Universe started in a state where the distance between any two points was zero, but there was still an infinite number of points. Then the distance between any two points started to increase.* **This means that there is no *special* point where the singularity happened.** All of the points in our Universe today were once part of the singularity. If no point is special, then no point can be the *centre* of the Universe.

The source of this misconception is probably the description of the big bang in popular science publications as *an explosion*. This seems appropriate because it's called *the Big Bang*. Well, this name is unfortunate. When most people imagine the Big Bang, they imagine a single point exploding

The Accelerated Motion 183

and expanding. But ... expanding into what? There's nothing outside of the Universe for the Universe to expand into.

Homogeneous and Isotropic

Homogeneous. Simply speaking a homogeneous material is something like a wall entirely made by identical bricks. Besides, such a wall does not have an edge, otherwise, it would not be homogeneous. So there is neither a start nor an end for this ideal wall.

Isotropic. Isotropy is essentially a point property, in fact, it is possible to notice that a certain object is isotropic with respect to a certain fixed point. To this purpose think about a star-shaped body: by definition, it is isotropic with respect to the centre of the star. Surely *homogeneous* and *isotropy with respect to a fixed point* are property which does not imply each other. On the other hand, isotropy with respect to each point implies certainly homogeneity.

Cosmological Principle

According to the *Cosmological Principle*, on large scales the Universe is isotropic with respect to each point, to this purpose think about the Cosmic Microwave Background Radiation which is surely both isotropic and homogeneous: the temperature differences are in the range of micro kelvin!

Universe Is Homogeneous so It Does Not Have an Edge or a Centre

Suppose the Universe is a boundary manifold: let x_0 belong to its boundary. Then the Universe would not be isotropic with respect to x_0, so globally it would not be homogeneous. On the other hand, the presence of a centre merely contradicts the hypothesis of homogeneity. So *any point in the Universe is its centre*.

Ball-Like - Positive-Curved Universe

Think about an ant moving straightforward on the skin of an apple: sooner or later it will return to its starting point. Surely the degree of freedom of such an ant is three: two spatial one plus the time. As a matter of fact moving inside the Universe there is three spatial degrees of freedom instead, so, under the hypothesis of *ball-like Universe* wherever is direct the three-velocity sooner or later everything will return at the starting point. Note the analogies and the differences: two spatial degree of freedom against three, something which can be thought, an apple's skin embedded in a 3−dimensional space against a three dimensional ball-shaped, embedded in ...nothing, since, as already stated, there is nothing outside of the Universe for the Universe to expand into.

Flat-Infinite Universe

A flat-infinite Universe may be thought of as a vectorial space, surely equipped with a metric (think about the Friedmann-Lemaître-R-W metric).

Consider than any point in the Universe and assume it as the Universe's centre: label it with O. Let $x \mapsto x/\sqrt{1 + \|x\|^2}$ be the map which apply each point $x = (x^1, x^2, x^3)$ in space onto the euclidean 3−open ball $B_O(1)$. This could be viewed as a manner to shrink an infinite extended object, namely a vectorial space onto a bounded one.

Flat-Finite Universe

No one could say if the Universe is finite or infinite, both in space and in time. The only thing to pay attention to is not to think of singularity as a mathematical point, as previously discussed.

Besides, do not forget the **negative-curved - hyperbolic Universe**!

And so, What's the Universe Like?

According to the actual measurement, it seems to be euclidean, which is flat. But, perhaps, men are making the same big mistake made by the pre-Columbian ...

9.13. Round Trip to the Edge of the Universe

This title wants to be deliberately provocative, just because no one knows whether the Universe is finite or infinite, flat or not.

Currently, something called dark energy is driving the accelerated expansion of the whole spacetime, namely the Universe is stretched in all directions.

Photon Moving on a Rubber Carpet

The title of this section is the paraphrase of the famous paradoxical problem of *the ant moving over a rubber rope*.

A photon starts to walk on a 3-space-dimensional rubber carpet (the Universe) at a constant speed of 299 792 458 m/s relative to the carpet it is walking on. At the same time, the carpet starts to stretch uniformly in all directions. The length of the carpet starts at $L_0 = 47$ billion light-years and then grows according to the law

$$L(t) = S(t) L_0$$

where $S(t)$ is the so-called *scale factor* and t denotes the cosmic time, namely the time which is beaten by *the Universe atomic clock*. Will the photon ever reach the end of the carpet?

It depends on the scale factor itself. In order to give a complete answer to this fascinating question, first of all, roll out a one-dimensional coordinate frame x parallel to the photon's direction of motion, with its origin coincident with the starting point of the photon. After having denoted by $x(t)$ the coordinate of the photon at time t, for every fixed value of t, the hypothesis of *uniformity* of the stretching undergone by the carpet means exactly that the following proportion holds, by applying a similarity argument between triangles:

$$x(t+dt) : x(t) = L(t+dt) : L(t)$$

which imply easily that

$$v_{x(t)}(t) : x(t) = v_{L(t)} : L(t)$$

i.e.

$$v_{x(t)}(t) = \frac{x(t)}{L(t)} v_{L(t)}.$$

Note that $v_{x(t)}(t)$ is not the speed of the photon, which is indeed $x'(t)$, but the speed at time t of the position of the Universe (the carpet) identified by the coordinate $x(t)$. The following proposition follows immediately:

Proposition 9.1 (Superluminal photons). *The speed of the photon is not constant, indeed exceed c by the addendum $\frac{x(t)}{L(t)}L'(t)$, namely:*

$$x'(t) = c + \frac{x(t)}{L(t)}L'(t) \qquad (9.52)$$

where, without any ambiguity $L'(t)$ denotes $v_{L(t)}$.

Note that it would be wrong to composite the two velocities c and $v_{x(t)}(t)$ according to Einstein's relativistic law of composition, as a matter of fact, although c is the speed of the photon relative to the Universe, $v_{x(t)}(t)$ is the speed of the *fixed* point *of the Universe* whose abscissa is $x(t)$ with respect to the initial position of the photon, considered as the origin.

Having (9.52) in mind let $y(t) = x/L$ so $x = L(t)y(t)$ and

$$x'(t) = L'(t)y(t) + L(t)y'(t).$$

It follows that:

$$L'(t)y(t) + L(t)y'(t) = c + y(t)L'(t) \Leftrightarrow y'(t) = \frac{c}{L(t)}$$

$$\Leftrightarrow x(t) = \int_0^t \frac{c}{L(t)}dt \cdot L(t).$$

So the photon will reach the end of the carpet if and only there exists $\bar{t}$ such that for all $t \geq \bar{t}$ it results $x(t) \geq L(t)$, if and only if $\int_0^t \frac{c}{L(t)}dt \geq 1$. Let's summarize this result in the following:

Proposition 9.2. *The photon will never reach the end of the carpet if and only if the following condition is satisfied:*

$$\forall \bar{t} \in \mathbb{R}^+ \quad \exists t < \bar{t} : \int_0^t \frac{c}{L(t)}dt < 1. \qquad (9.53)$$

The "Hubble's Law" and the Expanding Universe

It was Edwin Hubble's 1929 paper - *A relation between distance and radial velocity among extra-galactic nebulae*[15] - that led to a turning point in our

---

[15]Hubble E. (1929) *A relation between distance and radial velocity among extra-galactic nebulae.* Proceedings of the National Academy of Sciences of the United States of America 15(3) : 168 – 173.

The Accelerated Motion 187

understanding of the Universe. In his short paper, Hubble presented the observational evidence for one of science's greatest discoveries - the expanding Universe. Hubble showed that galaxies are *receding* away from us with a velocity that is proportional to their distance from us: more distant galaxies recede faster than nearby galaxies.

There is a linear relation between any galaxy velocity and its distance, which, **fixed a one-dimensional coordinate frame with the origin coincident with the Earth** (or, under the hypothesis of infinite Universe, in each other point in the Universe) is the following:

$$v(x) = H_o x. \qquad (9.54)$$

This is known as the *Hubble's law*, where H_o, the *Hubble Constant*, represents the constant rate of cosmic expansion caused by the stretching of spacetime itself. Although the expansion rate is constant in all directions at any given time, this rate changes with time throughout the life of the Universe. When expressed as a function of the cosmic time, $H = H(t)$ is known as the *Hubble Parameter*. The expansion rate at the present time, H_o, is about (70 ± 2) km/s/Mpc.

After a time dt a point in the Universe whose abscissa is x has receded by a quantity dx quantified in

$$dx = v(x)\,dt = H_o x\,dt \qquad H_0 = (70 \pm 2) \text{ km/s/Mpc}$$

which implies

$$\int \frac{dx}{x} = \int H_o\,dt \Leftrightarrow \ln|x| = H_o t + k \Leftrightarrow x(t) = k e^{H_o t}$$

Now set $t = 0$ so that $x(0) = L_0$ hence $x(t) = L_0 e^{H_o t}$.

Reaching the Edge of a Finite-Flat Universe

Suppose now that the Universe is flat and finite. With the same notations introduced in § 9.13., by using the same frame of coordinates prospected in § 9.13., let $L = L(t)$ denotes the overall length of the Universe at time t so that, standing the arbitrariness of x, by considering $x(t) = L(t)$ it results

$$L(t) = L_0 e^{H_o t}.$$

Consider now a mass-less body, namely a body whose speed is necessarily the maximum ever reachable, c, for example a photon, and invoke the proposition 9.53 with $L(t)$ above and set $L_0 = 47$ billion light years, which corresponds to the most distant object it is possible to see now. Is it possible for the photon to reach the point of the Universe whose abscissa is $L(t)$? In virtue of proposition 9.53 it is *not* possible if and only if for every fixed $\bar{t}$ there exists $t < \bar{t}$ (here expressed in Earth years) such that

$$\int_0^t \frac{c}{47 \cdot 10^9 \, c \, e^{H_o t}} dt < 1.$$

On the other hand it results that

$$\int_0^t \frac{c}{47 \cdot 10^9 \, c \, e^{H_o t}} dt = \frac{1}{-47 \cdot 10^9 \, H_o} \int_0^t -H_o \, e^{-H_o t} dt \qquad (9.55)$$

$$= \frac{1}{47 \cdot 10^9 \, H_o} \left(1 - e^{-H_o t}\right) \qquad (9.56)$$

which is less than 1 if and only if

$$1 - e^{-H_o t} < 47 \cdot 10^9 \, H_o$$

where this last inequality is obviously true for every t. So, if the Universe was finished it would be impossible to get out, like inside a prison.

... Unless to be able to construct an Einstein-Rosen's bridge or something which resembles an Alcubierre's warp-drive spaceship ...

Sky Will Be Forever Black

(i) Don't make any hypothesis regarding our Universe, (ii) think about yourself in a far-future era in which every star inside a 3−ball centred in your position whose radius is big enough (see further) has finished burning its nuclear fuel, and (iii) consider a photon coming from a faraway star directed towards your spatial position. Standing the hypothesis above, roll out a one-dimensional coordinate frame x parallel to the photon's direction of motion, with its origin coincident with the starting point of the photon. Let's denote by L_0 the distance at cosmic time $t = 0$ between the position in which the photon has been created and your position. Going through the topic of the § 9.13., after a time dt the point in the Universe whose abscissa is x has receded by a quantity $dx = H x dt$ so that at the generic time t the position of $x(t) = L_0 e^{Ht}$. Set $L = x$ and invoke proposition 9.53: is it true that for every $\bar{t}$ there exists $t < \bar{t}$ such that $\int_0^t \frac{c}{L_0 e^{Ht}} dt < 1$? By applying the same argument as in (9.55) it results that surely for every t (9.53) holds and so the sky will be forever black.

Note that here H is intended to denote the value of the Hubble constant in the far future prospected in (ii).

Chapter 10

The Accelerated Twin Paradox

Everyone who has faced Relativity in high school is familiar with the famous twin paradox. This chapter aims to study it in depth in its accelerated version, using the equations obtained in the previous chapter. The spacetime graphs of the twin's journey in the two different reference systems are also presented. Finally, a problem is solved with the aim of making the topic even more intuitive and interesting.

Keywords: Twin paradox, Speed-time graphs

10.1. The Original Twin Paradox

Time dilation equation 5.6 gives rise to a well-known relativistic argument, the so-called *Twin Paradox*[1]. The Twin Paradox is perhaps one of the most popular consequences of Einstein's theory of relativity.

---
[1] Initially called the Clock Paradox, it was renamed Twin Paradox in 1911 by French physicist *Paul Langevin*.

It is an ideal experiment aimed at illustrating how some aspects of Einstein's theory are contrary to common sense, but find an explanation in the context of the theory itself.

In fact the term *paradox* itself derives from the Greek adjective "παραδθξθς", which generally designates everything that overwhelms and contradicts "δoξα", in its most current meaning of "common opinion".

Explaining the Paradox

There are two twins, $\mathcal{A}$ and $\mathcal{B}$. One day the two separate: $\mathcal{A}$ stays on Earth while $\mathcal{B}$ gets on a spaceship and sets off on a space trip, which will keep him away from home for a few years (for his wristle clock). It is necessary to imagine that $\mathcal{B}$ moves with respect to $\mathcal{A}$ of uniform rectilinear motion. So there are two observers in two inertial reference frames.

Assuming that interstellar travel can be accomplished at speeds close to the speed of light, the theory predicts that, upon returning to earth, the "traveller" twin $\mathcal{B}$ has aged much less than the "terrestrial" one $\mathcal{A}$, in $\mathcal{A}$'s reference frame (see Figure 10.1).

In the eyes of the travelling twin, however, it is the Earth that is moving away so that the clocks on Earth will be slower and, from his point of view, it is the twin on Earth who will age less (see Figure 10.2).

But they can't both be right.

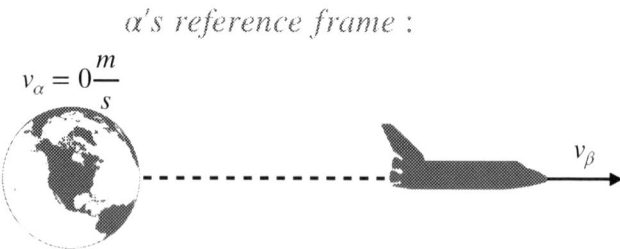

Figure 10.1.

The aspect that may perhaps seem paradoxical in the story of the twins is the apparent symmetry of the system: by choosing the spaceship as a reference system, it is the earth that moves away or approaches at speeds close to those of light.

The Accelerated Twin Paradox

β's reference frame :

Figure 10.2.

So why at the end of the journey is there a difference between the times measured by the twins? The solution is very simple: the two reference frames, the Earth and the spaceship, are not equivalent:

- The travelling twin $\mathcal{B}$, in fact, is not in an inertial reference system for the entire journey, the spaceship is subjected to acceleration at departure and arrival and, to an even greater extent, when she reverses course to return to Earth.

- The twin on Earth, on the other hand, is not affected by the forces associated with changes in speed and direction that the twin experiences while travelling. The twins are therefore distinguishable and their experiences are not symmetrical.

This implies that the special theory of relativity (which does not contemplate accelerations) cannot, therefore, be applied to the travelling twin. In any case, if the periods of acceleration are assumed to constitute only a small part of the overall journey, the special theory of relativity can be applied to give at least an indication of what happens: the conclusion is that the travelling twin will come back younger than the one left on Earth.

During the round trip at a constant speed, which constitutes most of the journey, both twins measure the same relative speed, but the proper length is measured by the twin on Earth, as he is at rest with respect to the points of departure and arrival. This implies, for special relativity, that the travelling twin measures a shorter distance, due to the phenomenon of the contraction of lengths. Therefore, while travelling at the same relative speed with respect to each other, from his point of view the travelling twin takes less time to cover a shorter distance and therefore may find it consistent that he will return home younger.

From the point of view of the twin on Earth, the heartbeat of the travelling twin is slower, so it is in accordance with the fact that the brother is younger when he returns.

An Everyday Paradox

Despite the apparent unfeasibility, the twin paradox has been experimentally verified. This is thanks to atomic clocks placed on board two planes flying in opposite directions with respect to the planet: the plane that travels in an easterly direction adds its speed to the rotation of the earth, therefore it travels faster than the one which travels westbound, and therefore must score a few fractions of a time shorter second. And so it was.

Another check experimental was instead performed in **1966** in a particle accelerator at CERN in Geneva. In this case, the travellers were muons, made to run through magnetic fields along circular paths with speed equal to 99.6 of the speed of light. It was found that, at their return, the muons were *younger*, because they had decayed more slowly than the muons at rest in the laboratory.

The only problem is that these experimental data were obtained by considering the effect of acceleration and deceleration on bodies. But these effects are not taken into consideration by the Special Relativity and by the twin paradox previously described. Therefore, in the following pages, the same paradox will be studied by exploiting the *relativistic equations of accelerated motion* obtained in the previous chapter 9.

10.2. Context and Data

In order to apply the equations obtained in chapter 9 to a practical example, it is first necessary to define the problem context and its data. They are shown below.

⋄ Two identical atomic clocks $\mathcal{A}$ and $\mathcal{B}$ are located in a point of the Universe far from any gravitational fields. For simplicity and to make the example more intuitive, the two clocks are initially placed on a "massless" space station.

◇ In the initial situation they are stationary and therefore have a speed equal to 0 m/s relative to each other.

◇ Both mark 00:00 when $\mathcal{B}$ embarks on a voyage away from gravitational fields aboard a spaceship which accelerates uniformly along a fixed direction for *one year* of time, according to $\mathcal{B}$.

◇ After the acceleration, $\mathcal{B}$ moves in a uniform rectilinear motion for *three years* (always according to $\mathcal{B}$) therefore, it decelerates uniformly for *one year* of time (according to $\mathcal{B}$) with the same deceleration in module.

◇ At the instant in which its velocity is cancelled, the spaceship reverses its motion and travels the line of moving backwards, with the same law reversed with respect to the outward journey.

The image shown in Figure 10.3 represents the situation at a certain Δt different from 0, in which the spaceship is moving straight towards a point in space, which will be its destination.

The aim of this chapter is to determine how much time $\mathcal{A}$ is ahead from $\mathcal{B}$, once $\mathcal{B}$ has returned to its starting point, the space station. It will thus be possible to demonstrate that there is nothing paradoxical in this example, seen from both the reference frames.

10.3. The Paradox No Longer Exists

First, it is useful to fix a one-dimensional Cartesian coordinate system oriented to the right, fixing its origin at the point where the spaceship leaves the space station, as shown in Figure 10.3.

The reason why it is possible to fix a one-dimensional coordinate system is that the spaceship, and therefore the space station, are located at a point in space far from gravitational fields. Furthermore, the effects of special relativity, in the absence of other forces, are present only along the axis of motion of a body.

Afterwards, a quick replay of the data can be useful for carrying out the problem.

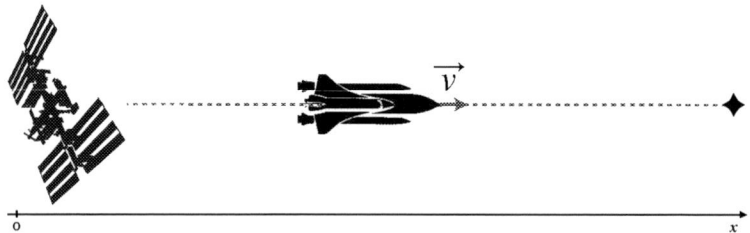

Figure 10.3.

Data Summary

$\mathcal{B}$'s reference frame:

Outward journey = $\begin{cases} \text{Uniformly accelerated motion} & \text{if } 0 \leq \tau \leq 1 \text{ year} \\ \text{Uniform rectilinear motion} & \text{if } 1 \text{ y.} \leq \tau \leq 4 \text{ y.} \\ \text{Uniformly decelerated motion} & \text{if } 4 \text{ y.} \leq \tau \leq 5 \text{ y.} \end{cases}$

Return journey = $\begin{cases} \text{Uniformly accelerated motion} & \text{if } 5 \text{ y.} \leq \tau \leq 6 \text{ y.} \\ \text{Uniform rectilinear motion} & \text{if } 6 \text{ y.} \leq \tau \leq 9 \text{ y.} \\ \text{Uniformly decelerated motion} & \text{if } 9 \text{ y.} \leq \tau \leq 10 \text{ y.} \end{cases}$

Target: determining $\Delta t - \Delta \tau$.

In order to better visualize the law of motion of $\mathcal{B}$, a speed-time graph of the outward journey in the reference of $\mathcal{B}$ is shown below in Figure 10.4. Starting from the equation 9.23, by deriving it with respect to τ it yields:

$$\frac{\partial s}{\partial \tau}(\tau) = c \cdot \tanh\left(\frac{A}{c}\tau\right). \tag{10.1}$$

Note that $A < 0$ imply $\frac{\partial s}{\partial \tau}(\tau) = c \cdot \tanh\left(-\frac{|A|}{c}\tau\right)$, whose graphic, compared with 10.1's one, is symmetric with respect to a suitable vertical axis. Below it is represented the graph of $\mathcal{B}$'s law of motion in the frame of the space station: By deriving the equation 9.10 with respect to t it gives (9.14),

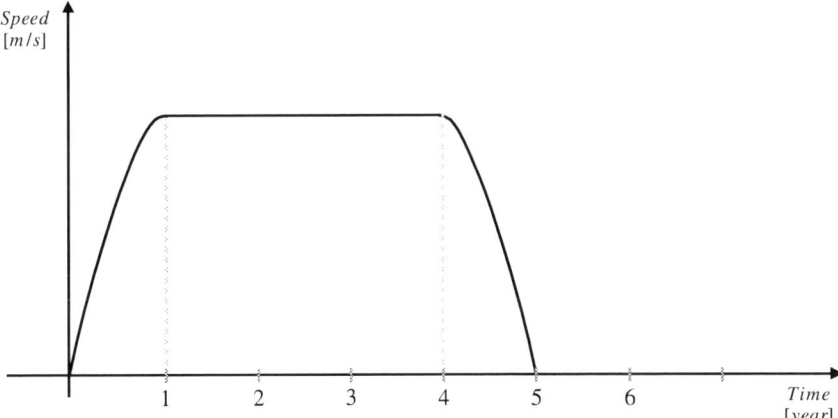

Figure 10.4. Graph of the speed $\frac{\partial s}{\partial \tau}$ against τ.

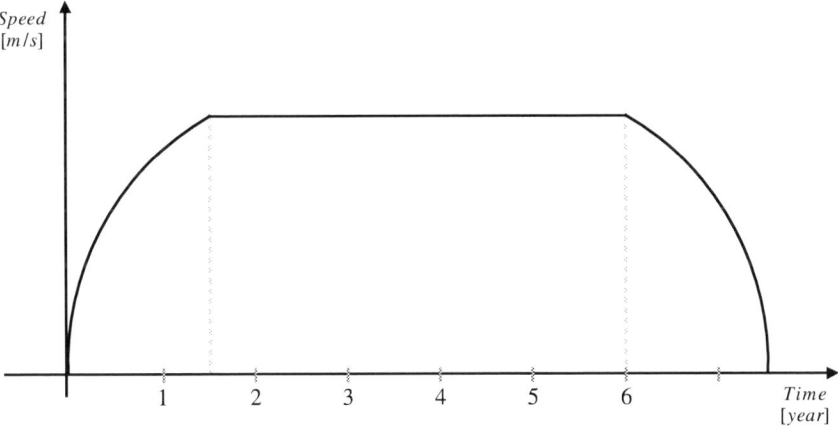

Figure 10.5. $\frac{\partial \sigma}{\partial t}$ against t.

whose graphs for both $A > 0$ and $A < 0$ are respectively the ascent and descent slopes of Figure 10.5. Now we are ready to proceed with the calculation of the time dilation.

First Phase: Acceleration

At the instant $t_0 = \tau_0 = 0\,\text{s}$ the spaceship turns on the engine and begins the acceleration phase until it reaches the speed v_{max} in a *non-proper* interval time Δt_1. Now, using the equations demonstrated in the previous chapter, it is possible to obtain:

$$\Delta t_1 = t(\tau = \Delta \tau_1) - t(\tau = 0) = \frac{c}{A_1}\sinh\left(\frac{A_1}{c}\Delta \tau_1 + \text{arsinh}\left(\frac{v_0 \gamma(v_0)}{c}\right)\right) - \frac{v_0 \gamma(v_0)}{A_1}$$

$$= \frac{v_0 \gamma(v_0)}{A_1}\cosh\left(\frac{A_1}{c}\Delta \tau_1\right) + \frac{c\gamma(v_0)}{A_1}\sinh\left(\frac{A_1}{c}\Delta \tau_1\right) - \frac{v_0 \gamma(v_0)}{A_1}$$

$$= 0 + \frac{c}{A_1}\sinh\left(\frac{A_1}{c}\Delta \tau_1\right) - 0 = \frac{c}{A_1}\sinh\left(\frac{A_1}{c}\Delta \tau_1\right)$$

since the initial speed of the spaceship is $v_0 = 0$ and so $\gamma(v_0) = 1$, where A_1 denotes the 4–acceleration of the first uniform *accelerated* phase (in particular this imply that $A_1 = |A_1|$). So finally

$$\boxed{\Delta t_1 = \frac{c}{|A_1|}\sinh\left(\frac{|A_1|}{c}\Delta \tau_1\right)} \qquad (10.2)$$

Second Phase: Uniform Rectilinear Motion

To calculate the time dilation in the uniform rectilinear motion it is sufficient to exploit the equation demonstrated by Einstein in 1905 through the ideal experiment of the *light clock* (see § 5.6). Therefore:

$$\Delta t_2 = \gamma \Delta \tau_2 = \frac{\Delta \tau_2}{\sqrt{1 - \left(\frac{v}{c}\right)^2}}$$

where v is the maximum speed reached in the first phase of uniform accelerated motion (whose initial speed is $v_0 = 0$):

$$v = v(\Delta t_1) = \frac{|A_1|\Delta t_1 + 0\gamma(0)}{\sqrt{1 + \left(\frac{|A_1|\Delta t_1 + 0\gamma(0)}{c}\right)^2}} = \frac{|A_1|\Delta t_1}{\sqrt{1 + \left(\frac{|A_1|\Delta t_1}{c}\right)^2}}$$

so that, standing the expression 10.2 for Δt_1

$$v = v(\Delta t_1) = \frac{|A_1|\frac{c}{|A_1|}\sinh\left(\frac{|A_1|}{c}\Delta \tau_1\right)}{\sqrt{1 + \left(\frac{|A_1|\frac{c}{|A_1|}\sinh\left(\frac{|A_1|}{c}\Delta \tau_1\right)}{c}\right)^2}} = \frac{c\sinh\left(\frac{|A_1|}{c}\Delta \tau_1\right)}{\cosh\left(\frac{|A_1|}{c}\Delta \tau_1\right)} = c\tanh\left(\frac{|A_1|}{c}\Delta \tau_1\right)$$

and finally it follows immediately:

$$\Delta t_2 = \Delta \tau_2 \cosh\left(\frac{|A_1|}{c}\Delta \tau_1\right) \qquad (10.3)$$

Third Phase: Deceleration

At this point, the spaceship passes to the deceleration phase that brings it up to zero speed and then reverses its direction to return to the space station.

For this phase, it is to be assumed either that the spaceship turns 180 degrees, or that it is also equipped with an engine in the front. In the first case, there would be a loss of energy and mass that is neglected.

The difference with respect to the first phase, that is the accelerated one, consists in the fact that the spaceship begins the deceleration phase with a speed $v_0 \neq 0$. If A_2 denotes the uniform 4−deceleration, by applying 9.26 with $-|A_2|$ instead of $|A_2|$ it follows

$$\Delta t_3 = \frac{c}{-|A_2|}\sinh\left(\frac{-|A_2|}{c}\Delta \tau_3 + \operatorname{arsinh}\left(\frac{v_0 \gamma(v_0)}{c}\right)\right) + \frac{v_0 \gamma(v_0)}{|A_2|}$$

where the above mentioned v_0 stands now for the speed achieved by the spaceship after the second phase, i.e. the speed achieved after the first accelerated phase:

$$v_0 = v(\Delta t_1) = c\tanh\left(\frac{|A_1|}{c}\Delta \tau_1\right)$$

so that

$$\gamma(v_0) = \cosh\left(\frac{|A_1|}{c}\Delta \tau_1\right) \rightsquigarrow v_0 \gamma(v_0) = c\sinh\left(\frac{|A_1|}{c}\Delta \tau_1\right)$$

and

$$\Delta t_3 = \frac{c}{-|A_2|}\sinh\left(\frac{|A_1|}{c}\Delta \tau_1 - \frac{|A_2|}{c}\Delta \tau_3\right) + \frac{c}{|A_2|}\sinh\left(\frac{|A_1|}{c}\Delta \tau_1\right).$$

Besides, note that the deceleration A_2 and the proper time $\Delta \tau_3$ (and so Δt_3) that such a deceleration lasts must be linked. As a matter of fact, in the accelerated frame the speed after $\Delta \tau_3$ must be equal to 0 (and so it must be in the inertial frame after a non-proper time equal to Δt_3). Correspondingly the following equations must hold:

$$v(\Delta t_3) = \frac{-|A_2|\Delta t_3 + c\sinh\left(\frac{|A_1|}{c}\Delta\tau_1\right)}{\sqrt{1 + \left(\frac{-|A_2|\Delta t_3 + c\sinh\left(\frac{|A_1|}{c}\Delta\tau_1\right)}{c}\right)^2}} = 0 \iff \boxed{\Delta t_3 = \frac{c}{|A_2|}\sinh\left(\frac{|A_1|}{c}\Delta\tau_1\right)}$$

(10.4)

and $\Delta\tau_3$ equals to

$$-\frac{c}{|A_2|}\text{arsinh}\left(\frac{-|A_2|\Delta t_3 + c\sinh\left(\frac{|A_1|}{c}\Delta\tau_1\right)}{c}\right) + \frac{c}{|A_2|}\text{arsinh}\left(\frac{c\sinh\left(\frac{|A_1|}{c}\Delta\tau_1\right)}{c}\right)$$

$$= -\frac{c}{|A_2|}\text{arsinh}\left(\frac{-|A_2|\Delta t_3 + c\sinh\left(\frac{|A_1|}{c}\Delta\tau_1\right)}{c}\right) + \frac{|A_1|}{|A_2|}\Delta\tau_1.$$

So that, putting them together:

$$\Delta\tau_3 = -\frac{c}{|A_2|}\text{arsinh}\left(\frac{-c\sinh\left(\frac{|A_1|}{c}\Delta\tau_1\right) + c\sinh\left(\frac{|A_1|}{c}\Delta\tau_1\right)}{c}\right) + \frac{|A_1|}{|A_2|}\Delta\tau_1$$

namely

$$\Delta\tau_3 = \frac{|A_1|}{|A_2|}\Delta\tau_1 \iff |A_2| = \frac{\Delta\tau_1}{\Delta\tau_3}|A_1| \qquad (10.5)$$

that is, after having assigned A_1, $\Delta\tau_1$ and $\Delta\tau_3$, the modulus of the 4−acceleration A_2 is obliged to be proportional to A_1, with constant of proportionality $\Delta\tau_1/\Delta\tau_3$.

If, as per hypothesis, $|A_1| = |A_2|$, if and only if $\Delta\tau_1 = \Delta\tau_3$, then invoking equation 10.4 it results:

$$\Delta t_3 = \frac{c}{|A_1|}\sinh\left(\frac{|A_1|}{c}\Delta\tau_1\right). \qquad (10.6)$$

Solution

In order to gain the solution, due to the symmetry of the problem, it's necessary to double the results obtained, since the spaceship makes not only the outward journey but also the return one. Therefore the solution is:

$$\Delta t_{\text{tot}} - \Delta \tau_{\text{tot}} = 2\left[\Delta t_1 + \Delta t_2 + \Delta t_3 - (\Delta \tau_1 + \Delta \tau_2 + \Delta \tau_3)\right] =$$

$$= 2\left[\frac{c}{|A_1|}\sinh\left(\frac{|A_1|}{c}\Delta\tau_1\right) + \Delta\tau_2 \cosh\left(\frac{|A_1|}{c}\Delta\tau_1\right)\right.$$

$$\left. + \frac{c}{|A_2|}\sinh\left(\frac{|A_1|}{c}\Delta\tau_1\right) - \Delta\tau_1 - \Delta\tau_2 - \Delta\tau_3\right]. \quad (10.7)$$

If, as per hypothesis, $|A_2| = |A_1|$ and $\Delta\tau_3 = \Delta\tau_1$, then more simply:

$$\Delta t_{\text{tot}} - \Delta \tau_{\text{tot}} = 2\left[\frac{c}{|A_1|}\sinh\left(\frac{|A_1|}{c}\Delta\tau_1\right)\right.$$

$$\left. + \Delta\tau_2 \cosh\left(\frac{|A_1|}{c}\Delta\tau_1\right) + \frac{c}{|A_1|}\sinh\left(\frac{|A_1|}{c}\Delta\tau_1\right) - 2\Delta\tau_1 - \Delta\tau_2\right]. \quad (10.8)$$

In order to compute the actual value of $\Delta t_{\text{tot}} - \Delta \tau_{\text{tot}}$, standing the value of the constant "speed of light in vacuum" $c = 299\,792\,458$ m/s, it is necessary to transform, for all i, $\Delta\tau_i$ in seconds, remembering that 1 *astronomical year* consists in 365.2422 days.

For example, if as per hypothesis $\Delta\tau_1 = \Delta\tau_3 = 1$ y, $\Delta\tau_2 = 3$ y, remembering the SEP and assuming $A_1 = 10\,\text{m/s}^2$ (so that, standing equation 10.5 $|A_2| = 10\,\text{m/s}^2$) then

$$\Delta t_{\text{tot}} - \Delta \tau_{\text{tot}} \approx 1616\,\text{days} \approx 4.42\,\text{y}.$$

Note that, under the hypothesis above, the speed reached after the first relativistic accelerated phase amounts to 78.28 % c, an exorbitant value for nowadays technologies.

Problem 8. *Think about a spaceship which uniformly accelerates with $A = 10\,\text{m/s}^2$ starting from a state of rest on Earth. What distance (indirectly) measured by itself will it cover, assuming that it continues to uniformly accelerate for 46 a (measured by the onboard clock)? Answer this question by using both Einstein's and Newton's equations.*

Solution 8. By using relativity, set $\tau = 10$ a in equation 9.24 so that

$$s(\tau) = \frac{c^2}{A}\ln\left(\cosh\left(\frac{A}{c}\tau\right)\right) \simeq \frac{(3\cdot 10^8)^2}{10}\cdot\ln\left(\cosh\left(\frac{10}{3\cdot 10^8}\cdot 46\cdot 3{,}15\cdot 10^7\right)\right) \simeq 4{,}3\cdot 10^{17}\,\text{m}$$

which corresponds to, according to equation 9.27:

$$\sigma(s) = \frac{c^2}{A}\left(e^{\frac{A}{c^2}s} - 1\right) \simeq \frac{(3 \cdot 10^8)^2}{10}\left(e^{\frac{10}{(3 \cdot 10^8)^2} \cdot 4,3 \cdot 10^{17}} - 1\right) \simeq 5,1 \cdot 10^{36} \text{ m}.$$

On the other hand, by using Newton's well-known mechanics:

$$s(\tau) = \frac{1}{2}a\tau^2 = \frac{1}{2} \cdot 10 \cdot \left(46 \cdot 3,15 \cdot 10^7\right)^2 \simeq 1,0 \cdot 10^{19} \text{ m}.$$

So, accordingly, to classical Newtonian mechanics, the spaceship will reach a proper distance much less than that it would reach by using relativity. On the other hand, such a less distance would be reached far exceeding the speed of light!

The results of the relativistic treatment can be understood, in the frame of the spaceship, due to the contraction of distances, which is the same thing, in the frame of the Earth, as the relativistic effect of dilation of time.

□

Part II
A Glimpse at General Relativity

Chapter 11

Gravitational Lensing and Proofs of General Relativity

This last chapter aims to take a detailed look at some of the aspects of General Relativity. It focuses on the concept of gravity and its representation, defining the concept of geodesics. The Christoffel symbols are then presented together with the Schwarzschild Metric. The maximal aging principle is studied in-depth and the Eddington and Briatore-Leschiutta experiments are analyzed.

Keywords: Gravity, Geodesic, Tangent Spaces, Christoffel symbols, Schwarzschild Metric, Maximal Aging, Eddington experiment, Briatore-Leschiutta experiment

In 1915 Albert Einstein publishes the *Theory of General Relativity*, the spark that will, later on, lead up to one of the biggest physical and philosophical revolutions of all time. Isaac Newton was the first to define gravity with *Philosophiae Naturalis Principia Mathematica*, published in 1687: he defined gravity as a force that draws massive objects towards each other, explaining how planetary systems and cosmology work; humans, for the first time, were able to understand the nature and the movement of planets, stars, any intergalactic medium. Since then the concept of gravity as a force became more and more eradicated in the culture, discerning the idea of a geocentric Universe, and even the heliocentric one. This idea, a century later Einstein's discovery, still holds up; this is the first reason why General Relativity is often considered difficult and counter-intuitive, because it goes against the common idea of gravity as a force, and many other wrong physical concepts eradicated in our culture.

The basic idea of General Relativity, in reality, is simple: what we call gravity, what causes the movement of planets, is a curvature of spacetime generated by the presence of planets. There's no such thing as gravity force, planets turn around the sun because the linear path they should follow is curved, and the cause of the curve is the mass of the sun. Any massive object that lives inside spacetime curves spacetime itself.

In Einstein's world, not only massive objects are subject to curvature on their path, so are massless objects such as waves and photons.

Gravity is given by the curvature of spacetime: the following graph, usually used to intuitively explain the concept of gravity, shows a geometrical representation of such spacetime's geometry.

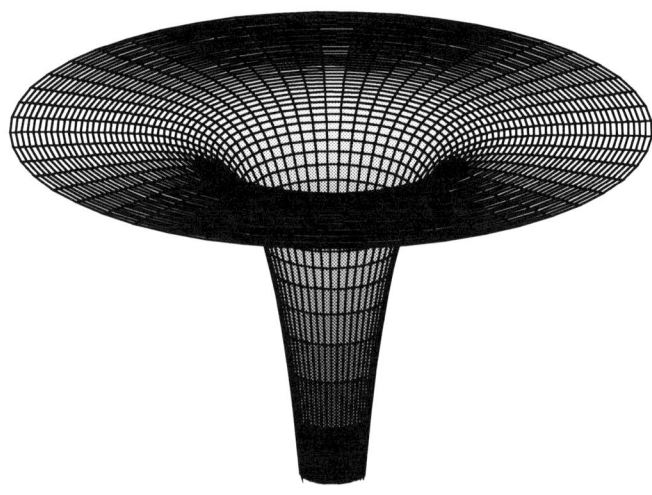

The flat surface is bent in the middle of it, as in the middle the presence of a massive object causes this. This representation, though, does not provide us with a strong comprehension of the phenomenon. In this representation, it seems like spacetime bends in a higher dimension, but this is not the case, even mathematically.

There's a new way to visualise general relativity today, more abstract, but clearer, where there's no need to add an unrealistic new dimension: gravity is given by the constant convergence movement of spacetime. The best way to think about this phenomenon is as if the geometric lines of spacetime, are constantly contracted towards massive objects: the magnitude of this convergence can be well visualized by a free-falling object, as the lines of space move the same way as the object does since the object belongs to spacetime and spacetime itself is made of mass, at least in part. The object remains motionless relative to the grid, but as the grid contracts, the object will fall. What allows the body to move, seen from a distant observer, is the time dimension of spacetime: the falling object moves because the lines of spacetime converge towards the planet as time passes. The rate of this convergence is always the same, as the mass of the planets is a fixed value. The curvature of spacetime appears to us as a constant contraction of the grid.

With this new representation in mind, it becomes clear that any object standing on the surface of a planet is constantly accelerating upwards since the only force applied to the body is the constraint force.

If time is really an element to be taken into consideration in the description of the motion of a body ... How is it possible to define such motion? The following section attempts to explain the concept of *geodesics*.

11.1. Geodesics

Now imagine observing an object with no motion. This object moves only through time since the space separation components are null.

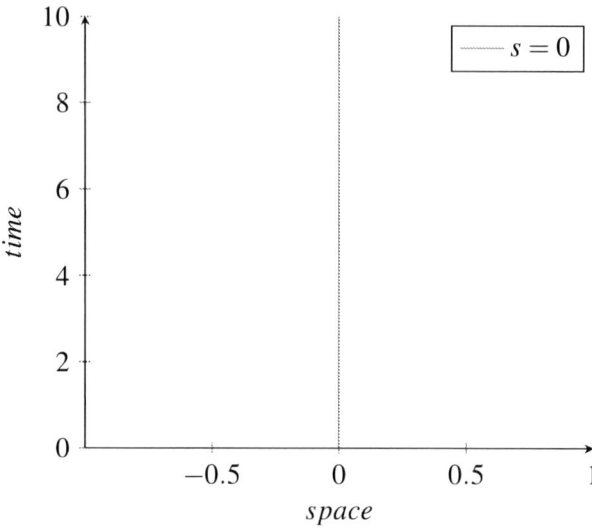

If the region of spacetime where the object lives becomes affected by gravity, for example by a planet, the lines of spacetime will eventually start to converge into the planet, carrying the body alongside with it, since mass not only lives inside spacetime, mass is part of spacetime. The object has no reason to move throw space: since gravity, as said earlier, is not a force, there is no force applied to the object. What happens is that the observer gets pushed towards the planet by the structure of spacetime itself, tracing a line called Geodesic.

A **Geodesic** is *the straightest possible path in a curved surface/space.* A **Geodesics curve** is defined as *a curve that has zero tangential acceleration when travelling along with it at a constant speed.*

Since the four-acceleration can be written as the derivative of the four-velocity with respect to proper time, the following equation holds:

$$\vec{A} = \frac{\partial \vec{V}}{\partial \tau} = \vec{0}. \tag{11.1}$$

In order to point out further, by definition of geodesic, the expression *when travelling along with it* means precisely that the time component has to be the one proper of the body "τ".

For any body, the natural movement follows a velocity vector that does not vary, *the natural movement is non-accelerating.*

Gravitational Lensing and Proofs of General Relativity

The four-velocity vector of a body can be written as the sum of its components multiplied by the basis vectors:

$$\vec{V} = V^0 \vec{e}_0 + V^1 \vec{e}_1 + V^2 \vec{e}_2 + V^3 \vec{e}_3 \qquad (11.2)$$

which - by using Einstein's standard convention - can be re-written as

$$\vec{V} = V^i \vec{e}_i.$$

Using this expression, it is possible to rewrite equation 11.1 as follows

$$\frac{\partial V^i \vec{e}_i}{\partial \tau} = \vec{0}$$

and indeed by Leibniz formula

$$\frac{\partial V^i}{\partial \tau} \vec{e}_i + V^i \frac{\partial \vec{e}_i}{\partial \tau} = \vec{0} \qquad (11.3)$$

in which the derivatives of the basis vector appear themselves. Basis vectors can indeed vary along the grid itself, if the grid is not flat.

Christoffel Symbols are the numbers that encode how the grid changes along each direction, allowing to define the "covariant derivative" D_τ along the line of Universe Σ: $\tau \mapsto \Sigma(\tau, \vec{x})$, where the spatial coordinate $\vec{x}$ is intended to be fixed:

$$\frac{\partial \vec{e}_i}{d\tau}(\tau_0) = D_\tau \vec{e}_i(\tau_0) = \Gamma^k_{ij}|_{\Sigma(\tau_0,\vec{x})} V^j|_{(\tau_0,\vec{x})} \vec{e}_k|_{\Sigma(\tau_0,\vec{x})} = \Gamma^k_{ij}|_{\Sigma(\tau_0,\vec{x})} \frac{\partial \Sigma^j}{\partial \tau}|_{(\tau_0,\vec{x})} \vec{e}_k|_{\Sigma(\tau_0,\vec{x})}$$

or simply

$$\frac{\partial \vec{e}_i}{\partial \tau} = D_\tau \vec{e}_i = \Gamma^k_{ij} V^j \vec{e}_k.$$

At this point the **Geodesic Equation** is defined by inserting the previous term in equation 11.3:

$$\frac{\partial V^i}{\partial \tau} \vec{e}_i + V^i \Gamma^k_{ij} V^j \vec{e}_k = \vec{0}$$

or, equivalently,

$$\frac{\partial V^k}{\partial \tau} \vec{e}_k + \Gamma^k_{ij} V^i V^j \vec{e}_k = \vec{0}$$

i.e.

$$\boxed{\frac{\partial V^k}{\partial \tau} + \Gamma^k_{ij} V^i V^j = 0} \qquad (11.4)$$

or, equivalently, by denoting with a "dot" the derivative with respect to proper time:

$$\ddot{\Sigma}^k + \Gamma^k_{ij}\dot{\Sigma}^i\dot{\Sigma}^j = 0. \tag{11.5}$$

This set of four equations allow us to calculate the rate of change of each component of the four-velocity as proper time passes:

$$\frac{\partial V^k}{\partial \tau} = -\Gamma^k_{ij}V^iV^j$$

Note that, anyway, by definition of geodesic, the derivative of the overall four-velocity must be zero:

$$\frac{\partial \vec{V}}{\partial \tau} = \vec{0}.$$

Exercise 19. *How is it possible to conciliate the fact that (i) the components of the vector $\vec{A}$ are possibly none equal to zero and (ii) the vector $\vec{A}$ equals to the null vector?*

11.2. Tangent Spaces

When an object follows a geodesic, falling into a source of gravity, for example a star or a planet, it passes from a flat region of spacetime to a curved one, as the curvature becomes bigger while getting closer to the planet. To mathematically analyse the properties of the body along its geodesic worldline it is necessary to consider that while moving it passes from a curved spacetime to another: the properties of space and time change along the way. The spacetime overall is thus covered by a system of charts possibly overlapping, such that, locally in each chart, the spacetime is inertial, flat, ruled by special relativity with Lorentz-Minkowsky norm. Besides, at each point in the curved surface of spacetime, it can be associated with a tangent space, which is just an abstract four-dimensional vectorial space. At each point corresponds a specific tangent space, with its own basis vectors.

To determine how basis vectors change along a certain path the **Metric Tensor** $g_{\mu\nu}$ is used, which enables to write the spacetime separation in a more general form as

$$(d\Sigma)^2 = g_{\mu\nu}dx^\mu dx^\nu \tag{11.6}$$

Gravitational Lensing and Proofs of General Relativity 211

where x^μ and x^ν are the local coordinates and μ, ν belong to $\{0,1,2,3\}$. Besides the equation above allows to write the norm of the 4−velocity vector by replacing the coordinate differences $d\Sigma$ by the four-velocity vector:

$$|\vec{V}|^2 = g(V^\mu \vec{e}_\mu, V^\nu \vec{e}_\nu) = V^\mu V^\nu g_{ij} dx^i \otimes dx^j (\vec{e}_\mu, \vec{e}_\nu) = V^\mu V^\nu g_{ij} \delta^i_\mu \delta^j_\nu$$
$$= V^\mu V^\nu g_{\mu\nu}. \quad (11.7)$$

The metric tensor describes the physical geometry of reality: it defines how coordinates change along the worldline, allowing us to calculate the Christoffel Symbols. It is obtained an equation that involves both the derivatives of the metric tensor - because Christoffel Symbols determines how basis vectors vary along the grid - and its inverse (g^{ij}), which is merely the inverse tensor with respect to (g_{ij}).

Christoffel Symbols Geometric Definition

$$\nabla_{\vec{e}_i} \vec{e}_j = \Gamma^k_{ij} e_k$$

Consider the *coordinate curve* σ_1 defined as

$$\sigma_1 : \tau \mapsto \tau \vec{e}_1$$

then $\sigma'_1(\tau) = \vec{e}_1(\tau) + \tau \frac{\partial \vec{e}_1}{\partial \tau}$ so that $\sigma'_1(0) = \vec{e}_1(0)$ and $\sigma_1(0) = O$. On the other hand, also the coordinate line x^1 (namely the line such that $x^2 = \hat{x}^2$ and $x^3 = \hat{x}^3$ are fixed) comes out of the point O along $\vec{e}_1(0)$ so that, for the unicity of the solution of a differential system with assigned initial data, locally around $\tau = 0$, σ_1 must coincide with the above mentioned coordinate curve.

The covariant derivative D along σ_1 in $\sigma_1(0) = O$ is

$$D_{\sigma_1(0)} \vec{e}_j(O) = \nabla_{\vec{e}_1(O)} \vec{e}_j(O) = \Gamma^k_{1j}(O) \vec{e}_k(O)$$

so that the Christoffel symbol Γ^k_{1j} quantifies the component along $\vec{e}_k$ of the vectorial variation of the basis vector $\vec{e}_j$ along the coordinate line x^1 (namely -along the direction $\vec{e}_1$), in other words, such a Christoffel symbol describes *how much of the change in the basis vector $\vec{e}_j$ with respect to the x^1 coordinate occurs in the direction of $\vec{e}_k$*.

Starting from the above mentioned definition the following expression of the Christoffel symbols can be derived in a conceptual-easy although articulated way.

Christoffel Symbols Metric Expression

$$\Gamma^k_{ij} = \frac{1}{2} g^{k\ell} \left(\frac{\partial g_{j\ell}}{\partial x^i} + \frac{\partial g_{i\ell}}{\partial x^j} - \frac{\partial g_{ij}}{\partial x^\ell} \right). \tag{11.8}$$

In a flat spacetime the **Minkowski metric** is used, in which the flat space metric is often denoted by the symbol $\eta_{\mu\nu}$ instead of $g_{\mu\nu}$. Using the so called *West Coast Convention* it results:

$$\eta = \begin{pmatrix} 1 & 0 & 0 & 0 \\ 0 & -1 & 0 & 0 \\ 0 & 0 & -1 & 0 \\ 0 & 0 & 0 & -1 \end{pmatrix}$$

Using this convention and homogenizing the spatial coordinates with respect to time, the spacetime separation is given by the following equation:

$$(d\tau)^2 = \eta_{\mu\nu} dx^\mu dx^\nu = dt^2 - \frac{1}{c^2} \left(dx^2 + dy^2 + dz^2 \right). \tag{11.9}$$

By applying HEI it is known that locally around each point of the Universe the curved spacetime is flat and modelled as the Lorentz-Minkowsky spacetime, ruled by special relativity. In particular, locally $g_{\mu\nu}$ coincides with $\eta_{\mu\nu}$, so that, locally, both $(d\Sigma)^2$ and $|\vec{V}|^2$ represent an invariant even for the metric $g_{\mu\nu}$, respectively the spacetime interval and 1 (or the speed of light c as well, depending on the homogenizing conventions). Baring in mind that the norm of the four-velocity is an invariant, it is possible to rewrite the expression 11.7 as

$$|\vec{V}|^2 = 1 = g_{\mu\nu} V^\mu V^\nu. \tag{11.10}$$

Note that by homogenizing with respect to *spaces* it gives instead

$$|\vec{V}|^2 = c^2 = g_{\mu\nu} V^\mu V^\nu.$$

11.3. Schwarzschild Metric

The Schwarzschild Metric was the first solution found for general relativity, a few months after its publication. It describes spacetime around a spherical mass, with no rotation and no charge. In this geometry, the metric tensor becomes

$$g_{\mu\nu} = \begin{pmatrix} 1 - \frac{2GM}{c^2 r} & 0 & 0 & 0 \\ 0 & -\left(1 - \frac{2GM}{c^2 r}\right)^{-1} & 0 & 0 \\ 0 & 0 & -1 & 0 \\ 0 & 0 & 0 & -\sin^2\theta \end{pmatrix} \quad (11.11)$$

With coordinates (homogenized with respect to times)

$$\left(x^0, x^1, x^2, x^3\right) = \left(t, \frac{r}{c}, \frac{r}{c}\theta, \frac{r}{c}\phi\right)$$

where r, θ and ϕ denotes respectively the distance from the center, the colatitude and the longitude.

The spacetime interval in spherical coordinates so becomes

$$d\tau^2 = \left(1 - \frac{2GM}{c^2 r}\right) dt^2 - \left(1 - \frac{2GM}{c^2 r}\right)^{-1} \left(\frac{dr}{c}\right)^2 - \left(\frac{r}{c}d\theta\right)^2 - \sin^2\theta \left(\frac{r}{c}d\phi\right)^2 \quad (11.12)$$

This equation, when talking about point masses such as a black hole, still holds up its value. Note that, knowing that **Schwarzschild radius** r_s is related to its mass M by

$$r_s = \frac{2GM}{c^2} \quad (11.13)$$

Equations 11.11 and 11.12 can be rewritten in function of the same r_s, as follows:

$$g_{\mu\nu} = \begin{pmatrix} 1 - \frac{r_s}{r} & 0 & 0 & 0 \\ 0 & -\left(1 - \frac{r_s}{r}\right)^{-1} & 0 & 0 \\ 0 & 0 & -r^2 & 0 \\ 0 & 0 & 0 & -r^2 \sin^2\theta \end{pmatrix}$$

$$d\tau^2 = \left(1 - \frac{r_s}{r}\right) dt^2 - \left(1 - \frac{r_s}{r}\right)^{-1} \frac{dr^2}{c^2} - \frac{r^2}{c^2}\left(d\theta^2 + \sin^2\theta d\phi^2\right).$$

The Schwarzschild solution only applies outside the massive object. In fact, the Schwarzschild radius r_s, for normal planets and stars is much smaller than the actual size of the object[1]. So, if $r \to r_s^+$ than $g_{tt} = g_{00} \to 0^+$ and $g_{rr} = g_{11} \to -\infty$.

The Schwarzschild radius is not defined as in 11.13, only mathematically: it has a deep physical meaning. The Schwarzschild radius represents the event horizon, the so-called *point of no return*: objects that pass the event horizon can no longer influence external observers, even light gets trapped inside by gravity.

An Historical Note

It follows a brief historical note useful for the next discussion:

- Einstein's Field Equation (1915);

- Schwarzschild's metric (1916): solving the Field Equation in the exterior region of a massive spherically symmetric body (e.g. a planet or a star);

- Time dilation in Schwarzschild's spacetime (e.g. Briatore-Leschiutta experiment);

- Geodesic associated to the Schwarzshild's metric: after having gained the Christoffel symbols of the Schwarzschild's metric, integrating the geodesic equation in order to find the trajectories described by free-falling, massive or massless particles;

11.4. Maximal Aging

To define the degree of aging of the different observers, it is necessary first to take into account that the contraction of space by gravity, also applies to time.

A free-falling object following a geodesic never moves through space - provided it was at rest at the beginning of the fall in a certain point of the spacetime - as space itself moves towards the planet as time goes by[2].

---

[1] So the maximum between the Schwarzschild radius and the physical radius of the planet or the star is typically the physical radius itself.

[2] Think about a free-falling elevator: it doesn't move throw space despite each point of it is moving!

Gravitational Lensing and Proofs of General Relativity   215

The time measured by the falling object though isn't the same as the one measured by an observer who does not move through space, living in a non-contracted region of spacetime.

The principle of **maximal aging** tells us that a particle that is not influenced by external forces follows the path that gives the largest possible proper time.

The places where time flows *faster*, where the absolute maximal aging occurs, are the flat regions of spacetime, where gravity does not affect the surface of spacetime (see below). As an object starts moving due to gravity along a geodesic it leaves the region of maximal aging, as it falls closer to the source of gravity (a planet, for example): the closer the object gets to the source of gravity, the more contracted the local region of spacetime is; the observer moving along a geodesic gradually starts to age less than the one located in a non-contracted region. If an observer sitting at a distance tending to infinite from a source of gravity shifts its position in space, he may enter a curved region of spacetime affected by gravity: even though in both states the observers could travel at the same speed through space, he wouldn't travel at the same speed through time. **The absolute maximal aging occurs in the region of spacetime not affected by gravity, in a state of rest.** *It follows the reason.* Consider two bodies one of which moves along a geodesic of the Schwarzschild's metric 11.12, the other being in a region of spacetime not affected by gravity, in a state of rest. Remember that, when using the equation of a metric, for example equation 11.12, it is necessary to consider two events: any two events happening on the observer moving along a geodesic, whose spacetime differential separation is given by $d\tau$, and its corresponding expression given by the observer at rest in the region not affected by gravity, with coordinates t, r, θ, ϕ. Taking into account the equation of the Schwarzschild metric 11.12, the time separation between the two events occurred along the geodesic is:

$$(d\tau)^2 =$$
$$= \left(1 - \frac{2GM}{c^2 r}\right)(dt)^2 - \left(1 - \frac{2GM}{c^2 r}\right)^{-1}\frac{(dr)^2}{c^2} - \frac{r^2}{c^2}\left((d\theta)^2 + \sin^2\theta\,(d\phi)^2\right)$$
$$\leq \left(1 - \frac{2GM}{c^2 r}\right)(dt)^2 \leq (dt)^2 \quad (11.14)$$

So $d\tau_{geo} < dt_{flat}$, namely the differential of the proper time, $d\tau$ measured by the observer moving along a geodesic is smaller than its corresponding interval of time measured by the observer at rest in a region not affected by gravity.

An observer standing on the surface of planet ages even less than the falling one, as he lives in the most contracted local region of spacetime, where time flows at the lowest possible rate. Not only that, he even moves through space in that local region, since the geometrical lines of space constantly fall inside the planet, as he stands still, obtaining an even bigger contraction of time, according to special relativity.

When a body moving along a geodesic undergoes an instantaneous acceleration, its spacetime path departs from the path of maximal aging along the geodesic it was following, and it ages more slowly. As a matter of fact, as stated in the paragraph above, the accelerated body will start moving both in time and in space, since he has moved from the previous geometric point of space to which he was previously attached. Locally, the moving accelerating body moves through space, recording less time separation between events, accordingly to equation 11.9. *In details*, if two coincident moving bodies following the same geodesic are considered, but then one of the two accelerates moving throw space[3], it is possible to analyse the local system using special relativity: this scenario is perfectly equivalent to all special relativity problems where there is a stationary observer (laboratory frame) and a moving observer (rocket frame). Another time, as discussed two paragraphs above, when using the equation of a metric, in this case, the special relativity metric, namely the Lorentz-Minkowsky's 11.9 one, it is necessary to consider two events: any two events happening on the moving observer are considered, the accelerating one, obtaining that the proper time $d\tau$ measured by this observer is smaller than the other one:

$$(d\tau)^2 = (dt)^2 - \frac{1}{c^2}(d\sigma_x)^2 - \frac{1}{c^2}(d\sigma_y)^2 - \frac{1}{c^2}(d\sigma_z)^2 < (dt)^2 \implies d\tau_{\text{acc}} < dt_{\text{geo}}$$
(11.15)

where dt, $d\sigma_x$, $d\sigma_y$ and $d\sigma_z$ are the separations between the two events measured from the stationary observer, namely, the non-accelerating (that is, moving along the geodesic) one.

Note that:

- in the argument used in (11.14) the Schwarzschild's metric was used instead of the Lorentz-Minkowsky's one since the comparison implicates two *distant* frames: the frame moving along the geodesic in the curved region governed by the Schwarzschild's metric and the

---

[3]Remember that who is moving along a geodesic never accelerates, as a matter of fact, the covariant derivative of its velocity with respect to the proper time is identically zero along the geodesic.

Gravitational Lensing and Proofs of General Relativity 217

frame at rest in the region not affected by gravity. On the contrary, in (11.15) the two frames coincide until one of them undergoes an instantaneous acceleration. So that, after a differential interval of time the two frames remain so near that EHI can be applied: this justifies the usage of the Lorentz-Minkowsky's metric.

- The accelerating observer plays the role of the uniform moving frame of special relativity, while the observer moving along the geodesic plays the role of the stationary frame of SR. For this purpose remark that the accelerating frame, since it is accelerating, departs from the geodesic line, so it moves in space; on the other hand, the frame moving along the geodesic is, by definition, a free-falling frame so it is equivalent to an inertial one. In § 11.4. is given a more detailed proof of the fact that such a frame really moves only in time and not in space, so its spatial velocity equals zero. Anyway, someone could infer that the accelerating frame does not belong to SR since gravity is involved, so that (11.15) is not applicable. For this purpose note that during the differential time dt the spatial separation of the accelerating frame from the geodesic one can only be a differential separation so that using this argument of locality the spacetime is flat and RS applies.

The Role of Acceleration

Equation 11.10 tells us that **every** frame of reference, inertial or not, moves *not only in space* but in *spacetime*, at the same constant speed, namely the speed of light in vacuum: c.

Relativistic acceleration, that is the 4−acceleration, is the unique responsible for the velocity distribution in the two *spatial and temporal containers*, which are in detail the γc and γv ones, so that, locally in each point of the spacetime, the Lorentz-Minkowsky norm of the 4−velocity (by homogenizing with respect to spaces) remains constantly c.

A frame of reference - say, improperly, *an observer* - could remain still sitting comfortably on an armchair in the desert, running by the sea or free-falling in a gravity field, but, anyway, he is moving in spacetime at constant speed c.

Consider for example a frame which is still in $P = (x^1, x^2, x^3)$ on the surface of a planet. Remember from above that t denotes the time measured by the MITCF comoving with the geodesic, passing through P at time $t = \tau = 0$, and τ is the time measured by the accelerating frame, still on the surface of the planet. So, as already observed, it results $d\tau_{acc} < dt_{geo}$, hence $\gamma \neq 1$ so $v \neq 0$; besides, by denoting with σ the distance of P measured by the geodesic frame, the *proper velocity*[4] $\gamma \vec{v}$ is equal to $\gamma \partial \vec{\sigma}/\partial t \neq 0$ and the 4−velocity is such that - by homogenizing with respect to spaces $|\vec{V}|^2 = \gamma^2 c^2 - \gamma^2 v^2 = \gamma^2 c^2 1/\gamma^2 = c^2$. Obviously, the last argument applies everywhere in the spacetime where acceleration is not zero.

Along a geodesic, since now the role of the accelerating frame is covered by the MITCF itself, it results $\tau = t$ so that $\gamma = \partial t/\partial \tau = 1$. Hence (i) $\gamma c = c$ and (ii) $v = 0$ so that $\gamma v = 0$, namely *the $\gamma \vec{v}$ container is empty*. It follows that *the γc container* reaches its minimum level if and only if the worldline is a geodesic.

In conclusion, a frame comoving with geodesic moves (with speed c) only along the time. Along any other worldline, the speed is distributed in both the containers (the difference between the square of them remains anyway always equal to c^2).

At last, note that maximality of ageing along a geodesic is consistent with the above discussion, in fact, $\gamma c = c\partial t/\partial \tau$ reaches its minimum value (equal to c) if and only if the differential time ∂t along a geodesic reaches its minimum value and the time $\partial \tau \leq \partial t$ is maximum. This verifies if and only if $\partial t = \partial \tau$, that is, the time along a geodesic reaches the maximum admissible time of an accelerating frame.

Summarizing: moving along a geodesic implies $\partial t_{geo} = \partial \tau$ so that

$$\partial t_{geo} = \partial \tau \Rightarrow \gamma = 1 \Rightarrow v \left(= \frac{\partial \sigma}{\partial t} \right) = 0 \Rightarrow \gamma v = 0.$$

On the other hand, if the frame is subjected to an acceleration then $\partial t_{geo} > \partial \tau \Rightarrow \gamma > 1 \Rightarrow \frac{\partial \sigma}{\partial t} = v \neq 0 \Rightarrow \gamma v \neq 0$. Hence a increases imply γc increases and so γv increases. So $a_2 > a_1 \Rightarrow \frac{\partial t_{geo}}{\partial \tau_2} > \frac{\partial t_{geo}}{\partial \tau_1} \Rightarrow \partial \tau_2 < \partial \tau_1$, in other words, if the acceleration increases then the speed of the geodesic time increases, that is, the time measured by the momentarily free falling frame increases with respect to proper time measured by the accelerated frame, so that in the accelerated frame, as the acceleration increases its value, the time flows more and more slowly. On the other hand, if the acceleration increases its

---

[4] Also known as *celerity*.

value then increases also $\frac{\partial \sigma}{\partial t} = v$, namely the velocity of the accelerated frame increases, the spatial motion increases its velocity. Note that $\partial \tau < \partial t_{geo} < \partial t_{flat}$. Besides, distances ∂s measured by the accelerated frame are more and more shrunk as acceleration increases its value. As a matter of fact, for any fixed $\partial \sigma_{geo}$, it results $\partial s = \partial \sigma_{geo}/\gamma$. In other words, *distance shrunk* is an effect of the velocity of the geodesic time with respect to the proper time (this was foreseeable since $\partial t = \gamma \partial \tau \Rightarrow \partial s = \partial \sigma/\gamma$).

In conclusion, as a frame $\mathcal{F}$ moves in spacetime, whenever it approaches a source of gravity its proper time will be contracted, or, equivalently, the geodesic time, which is the time measured by the momentarily comoving inertial frame, will be dilated with respect to the proper time measured by $\mathcal{F}$. Correspondingly, the *grid velocity* will undergo an increment. Conversely, as $\mathcal{F}$ gets further from the source of gravity it will expertise a less effect of contractions in time and correspondingly a decrease in the grid velocity.

11.5. The Eddington Experiment

In Einstein's theory of general relativity, any object path would get diverted into a geodesic, following the curvature of spacetime. In reality, the first consideration of gravitational deflection of light was published by Johann Georg von Soldner in 1801 but his predictions were much smaller than the ones of Einstein. Light and waves should be no different. This means that if distant stars are observed through a telescope on Earth, while the sun is in front of them, their light, deflected by the sun's gravity, would make them appear slightly out of position. If Einstein was right, then such phenomenon could have been seen during a solar eclipse, as the visible star's light beams, passing through the sun's contracted regions of spacetime, would appear visible to us. Confronting them with a picture taken at night, in a moment where those light beams paths won't be affected by gravity, they should result shifted in place.

Einstein's work though landed in the hands of astronomers, such as Charles Dillon Perrine, who made the first attempt at the 1912 solar eclipse in Brazil but failed due to the presence of clouds, which obscured star locations necessary to test Einstein's theory.

Two years later, Erwin Finlay-Freundlich tried again in the Solar eclipse of August 21, 1914; unfortunately, proving the experiment turned out to much harder than expected: World War I broke out and Freundlich and his equipment were interned in Russia, captured on the train they were travelling on by Russian police, before they could even set foot on the country, unable to carry out the necessary measurements. They will later be released in exchange for other prisoners.

The first successful result came during the total solar eclipse of 29 May 1919, where Arthur Eddington, a British astrophysicist, with the help of Frank Watson Dyson. Planning began in 1917, and in 1919 two expeditions departed England: one, led by Eddington, went in West Africa, to the islands of Principe, and the other headed to Brazil, in Sobral. Both locations were in the path of the eclipse and had favourable climates. Each group travelled with powerful photographic telescopes that could record detailed photos of the sky onto glass plates.

Back in England, Eddington compared the position of the bright stars of the Hyades from the eclipse plate with another of the night sky, using a machine that can take microscopic measurements within photos. The comparison revealed that the stars had shifted during the eclipse by roughly the amount Einstein predicted.

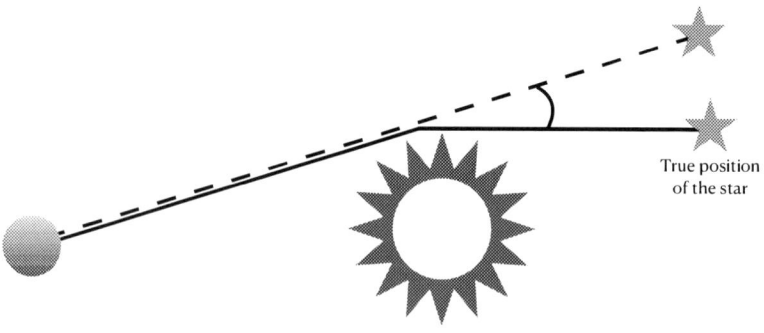

Figure 11.1.

The Eddington-Dyson experiment became the first proof of general relativity.

Gravitational Lensing and Proofs of General Relativity 221

11.6. The Briatore-Leschiutta Experiment

The Briatore-Leschiutta experiment from 1975 represents one of the clearest evidence of how gravity affects spacetime, as Einstein predicted in his theory. As it will be seen, the Schwarzschild metric is the easiest way to approach this model.

The experiment consisted in carrying two identical clocks in two different places, with different gravitational variables acting on them: one of the clocks was placed in Turin (clock 1), at the "Galileo Ferraris" institute, while the other lay at an altitude of 3250 m with respect to Turin, on the Plateau Rosà (clock 2), along the Matterhorn mountain. Turin's clock emitted an initial light type signal of initial synchronism and after 68 days it emitted a final signal of final synchronism, ending the experiment. What it is important to look for is the time recorded by the two different clocks: if general relativity is right, the two clocks should live in two regions of spacetime differently affected by gravity, and they should therefore record two different time intervals. Briatore and Leschiutta recorded a difference of 2.4 μs between the two clocks, an outcome very close to the previously calculated prediction, according to general relativity.

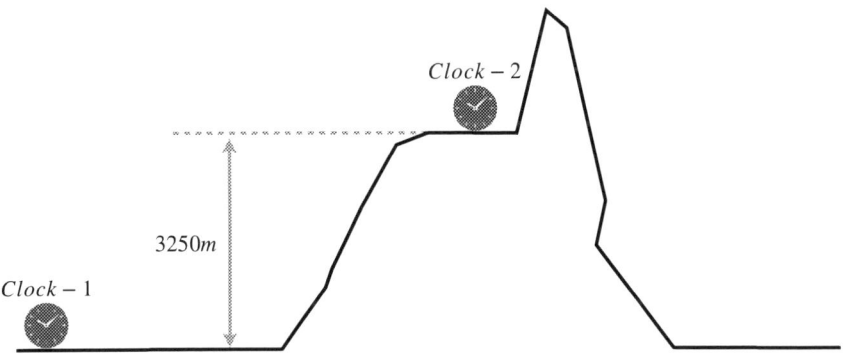

Figure 11.2. Problem visualization.

Solving the Problem: A Prediction of the Result

To mathematically calculate a solution of the model, the Schwarzschild metric will be used, since it describes how spacetime is affected by spherical masses, such as planet earth.

It will be assumed that the mass of the earth is the same for both frames of the clock since the most impacting variable for the two different positions is the distance from the centre of the earth.

Problem 9. (Briatore-Leschiutta experiment) *Clock* 1 *is located at sea level, clock* 2 *is located at* 3250 m *over sea level. Clock* 1 *records a time of* 68 days *(namely* $\Delta\tau_1 = 5\,875\,200.000\,000\,00$ s*).*

Calculate the time interval measured by clock 2*, knowing that earth's radius is approximately* 6 378 388 m *and that earth's mass is* $5.972\,190 \times 10^{24}$ kg.

Solution 9. Firstly, system containing the two equations of the two spacetime separations in Schwarzschild metric is wrote:

$$\begin{cases} ds_1{}^2 = c^2 d\tau_1{}^2 = \left(1 - \dfrac{2GM}{c^2 r_1}\right) c^2 dt^2 - \left(1 - \dfrac{2GM}{c^2 r_1}\right)^{-1} dr^2 - r_1{}^2 \left(d\theta_1^2 + \sin^2\theta_1 d\phi_1^2\right) \\ ds_2{}^2 = c^2 d\tau_2{}^2 = \left(1 - \dfrac{2GM}{c^2 r_2}\right) c^2 dt^2 - \left(1 - \dfrac{2GM}{c^2 r_2}\right)^{-1} dr^2 - r_2{}^2 \left(d\theta_2^2 + \sin^2\theta_2 d\phi_2^2\right) \end{cases}$$

Which, since the two rays are fixed and the slight distortion of space due to rotation is not considered, can be reduced to

$$\begin{cases} ds_1{}^2 = c^2 d\tau_1{}^2 = \left(1 - \dfrac{2GM}{c^2 r_1}\right) c^2 dt^2 \\ ds_2{}^2 = c^2 d\tau_2{}^2 = \left(1 - \dfrac{2GM}{c^2 r_2}\right) c^2 dt^2 \end{cases}$$

where r_1 and r_2 assume the following values:

$$r_1 = 6\,378\,388 \text{ m}$$

and

$$r_2 = 6\,378\,388 \text{ m} + 3250 \text{ m} = 6\,381\,638 \text{ m}.$$

Gravitational Lensing and Proofs of General Relativity

A Gaussian reduction is now applied, obtaining a relation between the two proper times:

$$\frac{c^2 d\tau_1^2}{c^2 d\tau_2^2} = \frac{\left(1 - \frac{2GM}{c^2 r_1}\right) c^2 dt^2}{\left(1 - \frac{2GM}{c^2 r_2}\right) c^2 dt^2} \Rightarrow \frac{d\tau_1^2}{d\tau_2^2} = \frac{\left(1 - \frac{2GM}{c^2 r_1}\right)}{\left(1 - \frac{2GM}{c^2 r_2}\right)} \Rightarrow d\tau_2^2 = \frac{\left(1 - \frac{2GM}{c^2 r_2}\right)}{\left(1 - \frac{2GM}{c^2 r_1}\right)} d\tau_1^2$$

$$\Rightarrow d\tau_2 = \sqrt{\frac{\left(1 - \frac{2GM}{c^2 r_2}\right)}{\left(1 - \frac{2GM}{c^2 r_1}\right)} d\tau_1^2} \Rightarrow d\tau_2 = d\tau_1 \sqrt{\frac{\left(1 - \frac{2GM}{c^2 r_2}\right)}{\left(1 - \frac{2GM}{c^2 r_1}\right)}}.$$

Since non-differential variables are being analyzed, it is necessary to transform all differentials d into Δ instead:

$$\Delta\tau_2 = \Delta\tau_1 \sqrt{\frac{\left(1 - \frac{2GM}{c^2 r_2}\right)}{\left(1 - \frac{2GM}{c^2 r_1}\right)}}$$

$$\Delta\tau_2 = 5\,875\,200.000\,000\,0\,\text{s} \sqrt{\frac{\left(1 - \frac{2 \cdot 6.6743\,\frac{\text{N} \cdot \text{m}^2}{\text{kg}^2} \cdot 5.972\,190 \times 10^{24}\,\text{kg}}{\left(299\,792\,458\,\frac{\text{m}}{\text{s}}\right)^2 6\,381\,638\,\text{m}}\right)}{\left(1 - \frac{2 \cdot 6.6743\,\frac{\text{N} \cdot \text{m}^2}{\text{kg}^2} \cdot 5.972\,190 \times 10^{24}\,\text{kg}}{\left(299\,792\,458\,\frac{\text{m}}{\text{s}}\right)^2 6\,378\,388\,\text{m}}\right)}}$$

$$= 5\,875\,200.000\,002\,1\,\text{s}$$

$$\Delta\tau_2 - \Delta\tau_1 = 5\,875\,200.000\,002\,1\,\text{s} - 5\,875\,200.000\,000\,0\,\text{s} = 2.1\,\mu\text{s}.$$

As it is possible to see, the result of our prediction model is $2.1\,\mu\text{s}$, notably close to the $2.4\,\mu\text{s}$ recorded in the actual experiment. □

This Paper-Sheet Is Too Wide!

Figure 11.3 shows the spacetime diagram in a *flat* chart of the situation prospected by Briatore and Leschiutta (see also Figure 11.2). The two diagonal sides of the plotted quadrilateral geometrically describe the lines of

Universe of the two rays of light both originated in Turin and addressed to Plateau Rosà: they are parallel since their gradients equal to 1. On the other hand, the two horizontal lines of the quadrilateral are also parallel since they are the line of Universe of Turin and the Plateau Rosà, which are - obviously - at a fixed height, namely 0 m and 3250 m. So the quadrilateral represented in Figure 11.3 is a parallelogram hence the two horizontal opposite sides must be congruent. On the other hand, as computed in the above paragraph, and shown by the experiment, they are not congruent! This apparent contradiction is solved by observing that the sheet of paper in which the parallelogram is represented is flat while in reality it is bent by gravity. In other words, the approximation of the spacetime around Turin and the Plateau Rosà by a flat spacetime is too coarse.

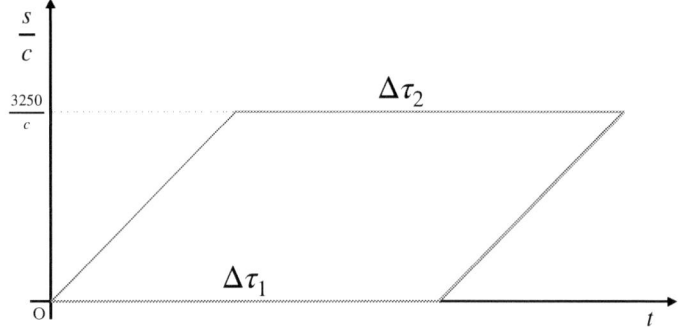

Figure 11.3.

The best way to physically understand why clock 2 ages more than clock 1 is to think about the newest model of representation of general relativity, presented in section 11.2.: first of all, both clocks are constantly accelerating, travelling through space, though clock 1 travels faster, since the geometric lines of space contract faster as something move closer to the massive object (earth). Moreover, both clocks live in curved regions of spacetime contracted by gravity, where time basis vectors are contracted (think about tangent spaces from 11.2.). Clock 1 lives in the most contracted region, since Earth's mass effects, are stronger closer to the centre of the planet, where density is stronger.

Part III
Conclusion

Chapter 12

Relativity in a Nutshell

> *The conclusion of this monograph could be partially summarized by the following citation due to Einstein:*
>
> **Short definition of Relativity**
> **There is no hitching post in the Universe**
> **— so far as we know**
>
> *By this definition, Einstein probably desired to point out the fact that, in the Universe, there is no point more important than the others, there is no absolute reference with respect to which refer all measurement of time and space, so far as we know.*

In 1908, Max Planck applied the term *Theory of Relativity* to describe the concepts carried out by Einstein's rising special theory. Instead of this title, probably Einstein would have preferred to refer to his creation with something like

Fundamental Invariant of Nature

which, instead, emphasizes what does *not* vary as the references change, such as mass, speed of light in vacuum, the spacetime interval and - more in general - all the physical quantities which are defined basing on tensors, such as four-position, -velocity, -momentum, -acceleration, -force. Also, the Christoffel symbols, on which the curvature of spacetime depends, are intrinsically defined, namely they do not depend on the local chart used to represent spacetime points.

In a nutshell, while eliminating the Newtonian notions of absolute time and space, Einstein based his theory *not on relativity* but *on what it is intrinsically defined*, which means on what does not vary, according to differentiable change in chart, namely, changing in reference frame. Einstein reduces time and space to a differentiable semi-Riemannian manifold, which is a curved four-dimensional geometric object together with a differentiable two-covariant tensor field, namely the metric tensor g, which is strongly linked to the *gravity* field itself. In a similar mathematical way Einstein addressed to gravity he tried to reduce electromagnetism and gravity together in terms of geometry, by using higher rank tensors. As a matter of fact, Einstein's ultimate paper was really dealt about this possibility.

Roughly speaking, on the one hand, it was possible to geometrize gravity thanks to the *equivalence principle*, which assures the possibility to locally delete the gravity, since it affects equivalently every massive or massiveless body; on the other hand, the *acceleration* due to the presence of an electric field acts equivalently on different charged body provided the quotient between their charges and masses are the same. So, it seemed to be needed exactly another one equivalence principle for each class of charges represented by the same quotient q/m.

But he failed to achieve this goal. In this respect, there is a nice story which regards the last person who spoke with Einstein. She was a nurse in Penn Princeton Medical Center who wheeled Einstein over to the bedside window to admire the view of the little round garden. She asked him if he thought God made the garden. He replied:

*Yes, God is both the gardener and the garden, I've spent my whole life just **trying to catch a glimpse** of him at his work.*

But, perhaps, it is intrinsically impossible *to look at God in action*, just as it is impossible to prove an indemonstrable proposition.

Part IV
Appendices

Appendix A

Hyperbolic Functions

In order to understand how a non-euclidean world works, it is necessary to introduce the hyperbolic functions.

A.1. Preamble

Hyperbolic functions, also called *hyperbolic trigonometric functions*, are defined in an analogous way to ordinary trigonometry. In fact, while the ordinary sine and cosine functions parametrize a circle, the hyperbolic sine and cosine parametrize *a hyperbola*.

Trigonometric Functions

Trigonometric functions are elementary functions, which relate angles to the sides of a right triangle. They are also called circular functions, since they can equivalently be defined as the lengths of various line segments from a unit circle.

Hyperbolic Functions

Hyperbolic functions are defined using the hyperbola $x^2 - y^2 = 1$ rather than the circle $x^2 + y^2 = 1$.
 The argument of hyperbolic functions is an **area**.

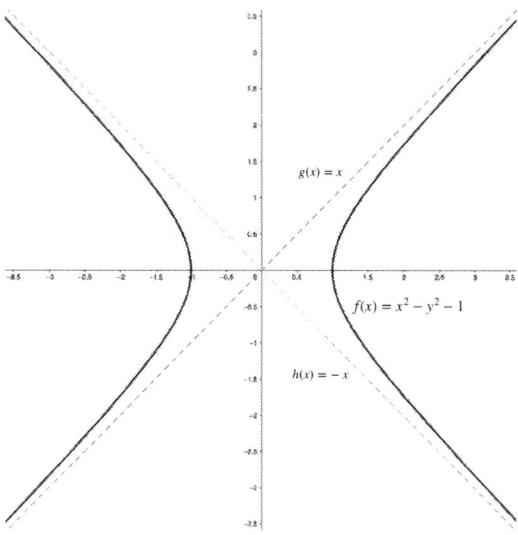

Figure A.1.

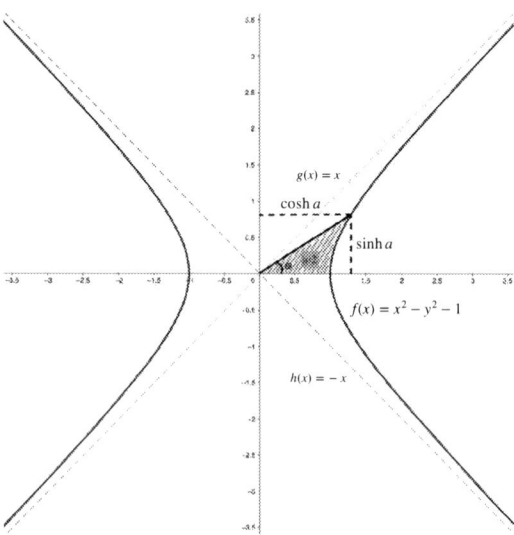

Figure A.2.

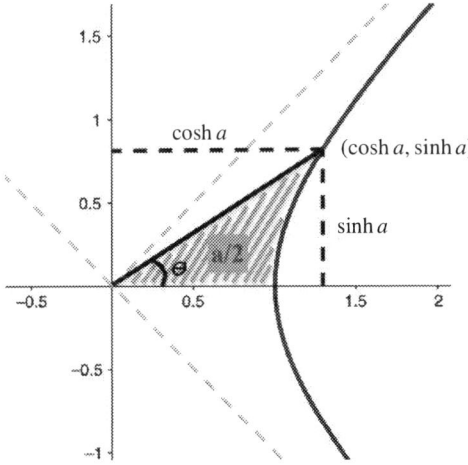

Figure A.3.

The points $(\cosh(a), \sinh(a))$ form the **right half of the unit hyperbola**. Therefore $\theta \neq 2a$ and

$$a \to \infty \iff \theta \to \frac{\pi}{4}.$$

Below are the definitions of the three hyperbolic functions

$$f(x) = \cosh(x), \quad f(x) = \sinh(x), \quad f(x) = \tanh(x).$$

A.2. Definitions

Definition of $f(x) = \cosh(x)$

The hyperbolic cosine function is defined for all real values of x by the relation:

$$\cosh(x) = \frac{e^x + e^{-x}}{2} \tag{A.1}$$

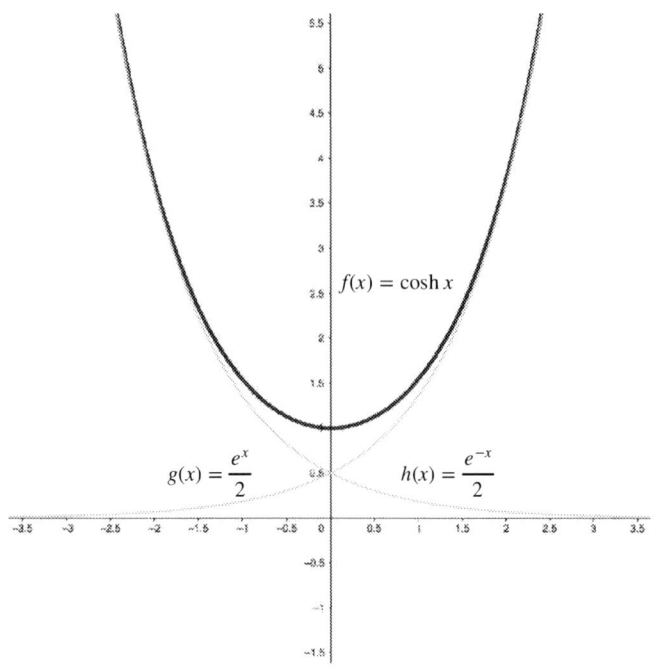

Figure A.4.

Key Points

1.

The hyperbolic cosine is an even function, since: $\cosh(x) = \cosh(-x)$.

2.

$\cosh(0) = 1$.

3.

$\forall x \in \mathbb{R}, \ \cosh(x) \geq 1$.

Hyperbolic Functions

4.

The graph $y = \cosh(x)$ is always above the graphs $y = \frac{e^x}{2}$ and $y = \frac{e^{-x}}{2}$.

Definition of $f(x) = \sinh(x)$

The hyperbolic sine function is defined for all real values of x by the relation:

$$\sinh(x) = \frac{e^x - e^{-x}}{2} \tag{A.2}$$

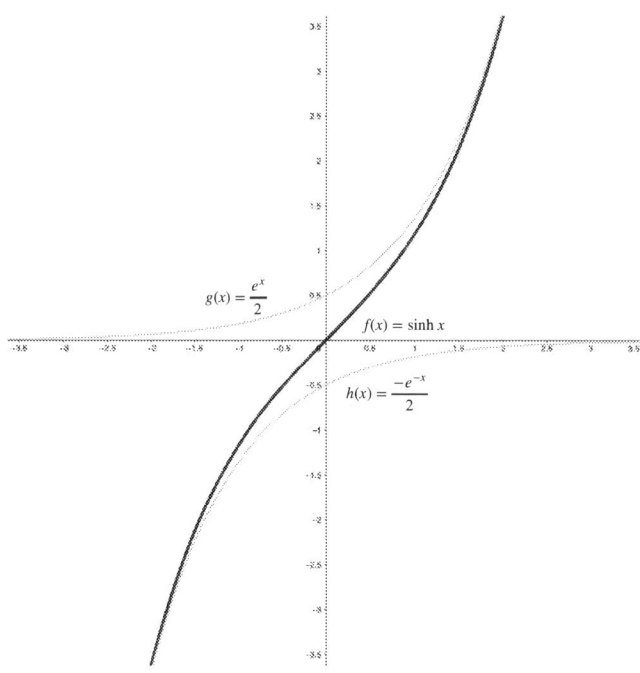

Figure A.5.

Key Points

1.

The hyperbolic sine is an odd function, indeed $\sinh(-x) = -\sinh(x)$ and so $\sinh(0) = 0$.

2.

The graph $y = \sinh(x)$ is bounded by the two curves $y = -\frac{1}{2}e^x$ and $y = \frac{1}{2}e^x$.

3.

$|\cosh(x) - \sinh(x)| = \cosh(x) - \sinh(x) = e^{-x}$.

Inverse Hyperbolic Functions

Both $\sinh(x)$ and $\cosh(x)$ are suitably made invertible; their inverse are denoted by $\mathrm{arsinh}(x)$ and $\mathrm{arcosh}(x)$ respectively.

A.3. Defining the Hyperbolic Functions

Let's start to differentiate the relation $x^2 - y^2 = 1$ taking into account the condition $x = 1 \Rightarrow y = 1$:

$$x^2 - y^2 = 1 \Leftrightarrow \begin{cases} x^2 - y^2 = 1 \\ x = 1 \wedge y = 0 \end{cases} \Leftrightarrow \begin{cases} x\,dx = y\,dy \\ x = 1 \wedge y = 0 \end{cases} \Leftrightarrow \begin{cases} x\frac{dx}{dt} = y\frac{dy}{dt} \\ x(0) = 1 \wedge y(0) = 0. \end{cases} \quad (A.3)$$

The differential system

$$\begin{cases} x'(t) = y(t) \\ y'(t) = x(t) \\ x(0) = 1 \wedge y(0) = 0 \end{cases} \quad (A.4)$$

surely imply (A.3). On the other hand (A.4) is surely equivalent to the second order differential equation $x'' - x = 0$ with initial data $x(0) = 1$,

$x'(0) = 0$ together with $y'' - y = 0$, $y(0) = 0$, $y'(0) = 1$. A straight computation of the characteristic polynomials of the above mentioned second order Cauchy's problems yields to the following solutions:

$$\begin{cases} x(t) = \frac{e^t + e^{-t}}{2} \\ y(t) = \frac{e^t - e^{-t}}{2}. \end{cases} \quad (A.5)$$

Hyperbolic functions are defined to be $x(t)$ and $y(t)$ above:

$$(\forall t \in \mathbb{R}) \quad (\cosh t, \sinh t) := \left(\frac{e^t + e^{-t}}{2}, \frac{e^t - e^t}{2} \right).$$

So $t \in \mathbb{R} \mapsto (\cosh t, \sinh t) \in \mathbb{R}^+ \times \mathbb{R}$ constitutes a parametrization for the right branch of the hyperbola $x^2 - y^2 = 1$. On the other hand $t \in \mathbb{R} \mapsto (-\cosh t, \sinh t) \in \mathbb{R}^- \times \mathbb{R}$ constitutes a parametrization for the left branch of the same hyperbola. Note that in both of them $t \in \mathbb{R}^+$ imply the parametrization of the upper half of each branch ($t \in \mathbb{R}^-$ parametrizes the lower half branches).

Standing the fact that $(\cosh t)' = \sinh t$ for every $t \in \mathbb{R}$, and $\sinh t > 0$ if and only if $t > 0$, it results that the mapping $t \mapsto \cosh(t)$ is invertible in $\mathbb{R}^+$. Similarly, sinh is invertible everywhere in its domain: $\mathbb{R}$. Then there are well defined functions $\sinh^{-1} : \mathbb{R} \to \mathbb{R}$ and $\cosh^{-1} : \mathbb{R}^+ \to \mathbb{R}^+$.

A.4. The Hyperbolic Tangent

The hyperbolic tangent function is defined for all real values of x by the relation:

$$\tanh x = \frac{\sinh x}{\cosh x} = \frac{e^x - e^{-x}}{e^x + e^{-x}} \quad (A.6)$$

Parity. The hyperbolic tangent is an odd function, indeed $\tanh(x) = -\tanh(x)$ and so $\tanh 0 = 0$.

Relationships. The graph of $\tanh x$ is very similar to the graph of $\arctan x$ as x approaches to zero. Indeed it's a notable fact that their Taylor's expansions centered in zero coincide up to order four: $\tanh x \simeq \arctan x \simeq$

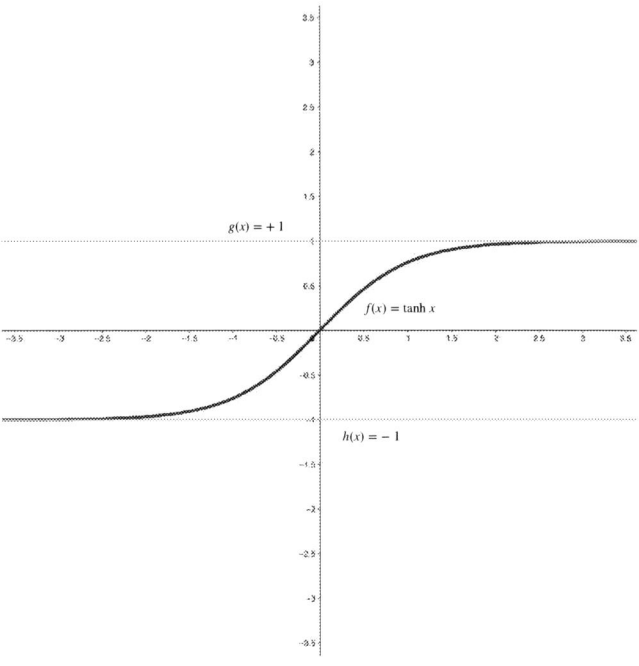

Figure A.6.

$x - x^3/3$ as x approaches to 0. This observation will result useful in the next section A.5. On the other hand it oscillates between ± 1 instead of $\pm\pi/2$.

Inverse function. Sometimes it could be useful to take into account the explicit value of the inverse function of the hyperbolic tangent, artanh, defined for all y such that $-1 < y < 1$:

$$\operatorname{artanh}(y) = \frac{1}{2}\ln\frac{1+y}{1-y} \qquad (A.7)$$

whose proof is a straightforward application of the exponential properties.

Hyperbolic Functions

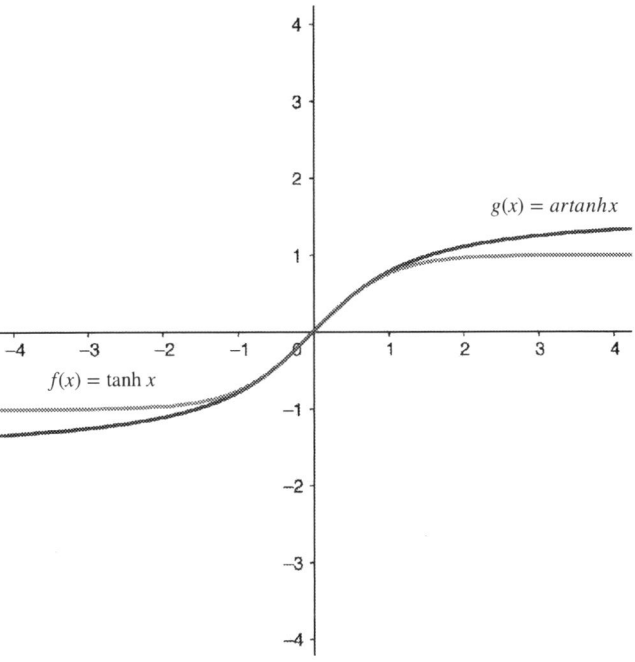

Figure A.7.

A.5. Full Geometric Interpretation

Trigonometric functions are closely linked with the so called trigonometric circumference. In details, they are concerned with angles, arcs, segments: triangles. In order to introduce hyperbolic functions, it's useful to put in evidence another way to deal with trigonometric functions, as shown below. The area A of the circular sector OAP is equal to:

$$A = \frac{\theta}{2} r^2$$

and, for the trigonometric circumference, in which $r = 1$

$$A = \frac{\theta}{2} \iff \theta = 2A.$$

So by writing for example $\sin \theta$, it is the same as $\sin(2A)$.

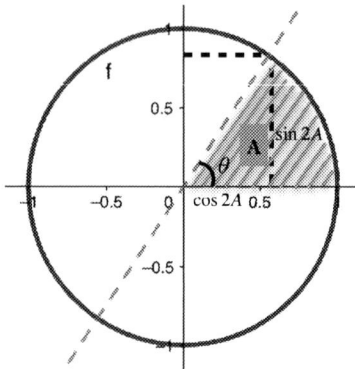

Figure A.8.

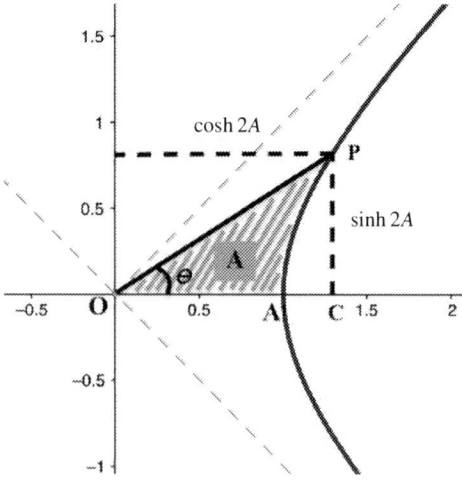

Figure A.9.

The area A of the hyperbolic sector OAP is equal to the difference between the area of the triangle OPC and the area of the plane region bounded

Hyperbolic Functions 241

by the arc of hyperbola *AP*, the *x* axis and *PC*. Therefore:

$$A = \frac{1}{2}x_p\sqrt{x_p^2 - 1} - \int_1^{x_p} \sqrt{x^2 - 1}\, dx. \tag{A.8}$$

In order to calculate the primitive of $\sqrt{x^2 - 1}$, after having noted that the graph of the function $x \mapsto \sqrt{x^2 - 1}$ consists in the two *upper* semi-branches of the hyperbola $x^2 - y^2 = 1$, then, substituting $x = \cosh t$ (hence, assuming $x \geq 1$):

$$\int \sqrt{x^2 - 1}\, dx = \int \sqrt{\sinh^2 t}\, d(\cosh t) = \int \sinh t \sinh t\, dt = \int \sinh^2 t\, dt. \tag{A.9}$$

To proceed in the computations, first of all note that (straight check)

$$\cosh(x+y) = \cosh(x)\cosh(y) + \sinh(y)\cosh(x)$$

and so

$$\cosh(2x) = \cosh^2(x) + \sinh^2(x) = 2\cosh^2(x) - 1 = 1 + 2\sinh^2(x),$$

in which the *fundamental hyperbolic relationship* has been involved. Then (A.9) becomes

$$\int \frac{\cosh(2t) - 1}{2}\, dt = \frac{\sinh(2t)}{4} - \frac{1}{2}t = \frac{\sinh(t)\cosh(t)}{2} - \frac{1}{2}t. \tag{A.10}$$

On the other hand, remembering that $x = \cosh(t)$ and so

$$(\forall x \geq 1) \qquad t = \cosh^{-1}(x) = \ln\left(x + \sqrt{x^2 - 1}\right)\, (\geq 0),$$

$t = \ln\left(x - \sqrt{x^2 - 1}\right) \leq 0$ parametrizes the lower right semi-branch, equation A.10 become

$$\frac{1}{2}\sinh\left(\ln\left(x + \sqrt{x^2 - 1}\right)\right)\cosh\left(\ln\left(x + \sqrt{x^2 - 1}\right)\right) - \frac{1}{2}\ln\left(x + \sqrt{x^2 - 1}\right)$$
$$= \frac{1}{2}x\sqrt{x^2 - 1} - \frac{1}{2}\ln\left(x + \sqrt{x^2 - 1}\right). \tag{A.11}$$

It is now possible to compute the integral in A.8:

$$\int_1^{x_p} \sqrt{x^2 - 1}\, dx = \frac{1}{2}x_p\sqrt{x_p^2 - 1} - \frac{1}{2}\ln\left(x_p + \sqrt{x_p^2 - 1}\right).$$

Finally (A.8) becomes:

$$A = \frac{1}{2}\ln\left(x_p + \sqrt{x_p^2 - 1}\right) = \frac{1}{2}\cosh^{-1}(x_p)$$

if and only if

$$\cosh^{-1}(x_p) = 2A.$$

From this point of view the inverse function of the hyperbolic cosine is also called arcosh, since it computes the area whose hyperbolic cosines is x_p.

Finally, as previously noted in A.4., for small values of their arguments the corresponding values of arctan and tanh almost coincides, and so, given

$$P = (\cosh(2A), \sinh(2A)),$$

if θ is sufficiently small (iff $A \ll 1$) then $\arctan(2A) \simeq \tanh(2A) = \tan\theta$. On the other hand, being both θ and $2A$ small, both tan and arctan coincides up to the second order's Taylor expansion centered in zero, so that

$$2A \simeq \theta$$

up to the second order, and the comparison with the trigonometric functions is now complete.

In order to gain the hyperbolic identities (addition, duplication) used in this section it could be useful to remember the so called *Osborn's rule*, aimed to convert a trigonometric identity *not involving calculus* into a corresponding hyperbolic one.

Osborne's rule: (a) Every occurrence regarding sine or cosine has to be replaced with the corresponding hyperbolic sine or cosine; (b) wherever one has a product of two sines, possibly with different arguments, the product of the two hyperbolic sines must be negated.

Bibliography

[Ein05] Einstein Albert, "Zur Elektrodynamik bewegter Körper. (German) [On the electrodynamics of moving bodies]". In: *Annalen der Physik* 322.10 (1905), pp. 891–921. DOI: http://dx.doi.org/10.1002/andp.19053221004.

[Gas10] Gasperini Maurizio, *Manuale di Relatività Ristretta*. Springer, 2010.

[Car19] Carol Sean M., *Spacetime and Geometry*. Cambridge University Press, 2019.

About the Authors

Riccardo Zancan
Energy Engineering Student
University of Pisa (UNIPI), Livorno (LI), Italy
Email: riccardo.zancan02@gmail.com

Riccardo Zancan was born in Abano Terme (PD), Italy, on June 28, 2002. He attended the Federigo Enriques scientific high school in Livorno and is currently a student of Energy Engineering at the University of Pisa. He loves sailing and is passionate about mathematics, physics, and culture in general.

Raul Tozzi
Professor of Mathematics and Physics
State Scientific High School "Federigo Enriques"
Regional School Office for Tuscany, Livorno (LI), Italy
Email: raul.tozzi@gmail.com

Raul Tozzi was born in Livorno, Italy, on November 7, 1981. He graduated in mathematics and currently teaches mathematics and physics. He is curious about sciences and enjoys working in multicultural environments.

Index

Aberration, 57
Aberration angle, 57
Accelerated Motion, 135
Accelerated twin paradox, 191
Aether, 30
Axiom, 35

Ball-like - positive-curved Universe, 183
Boundary analysis, 157
Bradley, 57
Briatore-Leschiutta experiment, 221

Car and garage paradox, 88
Causality, 96
Christoffel Symbols, 209
Christoffel symbols in metric expression, 212
Chronological order, 93
Chronotope, 104
Composition of velocities, 65
Correspondence Principle, 39
Cosmological Principle, 46, 183
Cosmology, 182
Crisis of classical physics, 29
Criticism to the contraction of distances, 55
Curvature of a ray of light, 161

Definitions, postulates and principles, 35
Deriving space contractions from L.T., 64
Deriving time dilation from L.T., 64
Deriving with respect to proper time, 107

Eddington experiment, 219
Edge of a finite-flat Universe, 187
Edge of the Universe, 184
Einstein's train paradox, 70
Energy, 111
Equation summary for accelerated motion, 156
Event, 36

Finite difference Lorentz equations, 62
First Postulate of Special Relativity, 39
Flat-finite Universe, 184
Flat-infinite Universe, 184
Four-Acceleration, 107, 138
Four-Acceleration in MITCF, 141
Four-Force, 114
Four-Momentum, 112
Four-Position, 104
Four-Velocity, 105
Frame of Coordinates, 36

Frame of Reference, 36
Galilean transformation, 37
Gamma factor, 52
General Covariance, 43
General Relativity Principle, 43
Geodesics, 207
Geodesics curve, 208
Graph of accelerated motion, 148
Gravitational lensing, 205
Gravity, 206

Heat, 112
Homogeneous, 183
Horizons, 168
Horizons in Rindler's coordinates, 176
Hubble's Law, 186
Hyperbolic functions, 231
Hyperbolic motion, 158

Indirect relativistic measures, 55
Invariant interval, 66
Inverse hyperbolic functions, 236
Isotropic, 183

Kronecker product, 27

Law, 35
laws of dynamics, 38
Light Clock, 47
Light speed, 40
Light's climb rate, 56
Light-meter, 41
Light-second, 41
Lorentz space-equation, 60
Lorentz time-equation, 61
Lorentz transformations, 59

Lorentz-Minkowsky's Flat Space-Time, 103
Lorentz-Minkowsky's Metric, 107

Matrix, 26
Maximal Aging, 214
Metre, 41
Michelson-Morley experiment, 30
Misconception about SR, 136
Momentarily Inertial Tangent Co-moving Frame (MITCF), 136
Moving away in opposite directions, 180

Negative-curved - hyperbolic Universe, 184

Perpendicular distances, 48
Photon moving on a rubber carpet, 185
Pinor, 26
Postulate, 35
Principle, 35
Principle of Causality, 96
Principle of Maximal Aging, 46
Proper length, 36
Proper time, 37
Properties of Lorentz equations, 63

Quoting two non-proper quantities, 109
Quoting two proper quantities, 108

Relative relativistic uniformly accelerated motions, 178
Rindler's coordinates, 163
Rindler's metric, 162

Scalars, 28
Scale factor, 185
Schwarzschild Metric, 212
Second Postulate of Special Relativity, 40
Simultaneity, 69
Singularity-single point, 182
Sky will be forever black, 188
Solved exercises, 115
Spacetime, 104
Special Covariance, 43
Special Relativity Principle, 42
Speed versus time graphs, 161
Strong Einstein Equivalence Principle, 44
Strong Equivalence Principle (SEP), 45

Tangent Space, 210
Tensor, 26
The Happiest Idea of my life (EHI), 44
Train paradox-light sensors' variation, 79
Trigonometric functions, 231
Twin Paradox, 191

Uniformly Accelerated Motion, 138
Universe Lines, 110
Universe's shape, 182

Vector, 25
Versor, 26

Weak Equivalence Principle, 43
West-Coast Convention, 107
Work, 112

Zeeman effect, 40

The Bible says the earth stands still, my dears
A fact which every learned doctor proves:
The Holy Father grabs it by the ears
And holds it hard and fast.
And yet it moves

– The Life of Galileo, Bertolt Brecht

*"**One**, remember to look up at the stars and not down at your feet.*
***Two**, never give up work*
Work gives you meaning and purpose and life is empty without it.
***Three**, if you are lucky enough to find love*
remember it is there and don't throw it away"

– Stephen Hawking